Cipriano Galindo, Juan-Antonio Fernández-Madrigal and Javier González

# Multiple Abstraction Hierarchies for Mobile Robot Operation in Large Environments

# Studies in Computational Intelligence, Volume 68

Editor-in-chief
Prof. Janusz Kacprzyk
Systems Research Institute
Polish Academy of Sciences
ul. Newelska 6
01-447 Warsaw
Poland
*E-mail:* kacprzyk@ibspan.waw.pl

Further volumes of this series
can be found on our homepage:
springer.com

Vol. 47. S. Sumathi, S. Esakkirajan
*Fundamentals of Relational Database Management Systems*, 2007
ISBN 978-3-540-48397-7

Vol. 48. H. Yoshida (Ed.)
*Advanced Computational Intelligence Paradigms in Healthcare*, 2007
ISBN 978-3-540-47523-1

Vol. 49. Keshav P. Dahal, Kay Chen Tan, Peter I. Cowling (Eds.)
*Evolutionary Scheduling*, 2007
ISBN 978-3-540-48582-7

Vol. 50. Nadia Nedjah, Leandro dos Santos Coelho, Luiza de Macedo Mourelle (Eds.)
*Mobile Robots: The Evolutionary Approach*, 2007
ISBN 978-3-540-49719-6

Vol. 51. Shengxiang Yang, Yew Soon Ong, Yaochu Jin Honda (Eds.)
*Evolutionary Computation in Dynamic and Uncertain Environment*, 2007
ISBN 978-3-540-49772-1

Vol. 52. Abraham Kandel, Horst Bunke, Mark Last (Eds.)
*Applied Graph Theory in Computer Vision and Pattern Recognition*, 2007
ISBN 978-3-540-68019-2

Vol. 53. Huajin Tang, Kay Chen Tan, Zhang Yi
*Neural Networks: Computational Models and Applications*, 2007
ISBN 978-3-540-69225-6

Vol. 54. Fernando G. Lobo, Cláudio F. Lima and Zbigniew Michalewicz (Eds.)
*Parameter Setting in Evolutionary Algorithms*, 2007
ISBN 978-3-540-69431-1

Vol. 55. Xianyi Zeng, Yi Li, Da Ruan and Ludovic Koehl (Eds.)
*Computational Textile*, 2007
ISBN 978-3-540-70656-4

Vol. 56. Akira Namatame, Satoshi Kurihara and Hideyuki Nakashima (Eds.)
*Emergent Intelligence of Networked Agents*, 2007
ISBN 978-3-540-71073-8

Vol. 57. Nadia Nedjah, Ajith Abraham and Luiza de Macedo Mourella (Eds.)
*Computational Intelligence in Information Assurance and Security*, 2007
ISBN 978-3-540-71077-6

Vol. 58. Jeng-Shyang Pan, Hsiang-Cheh Huang, Lakhmi C. Jain and Wai-Chi Fang (Eds.)
*Intelligent Multimedia Data Hiding*, 2007
ISBN 978-3-540-71168-1

Vol. 59. Andrzej P. Wierzbicki and Yoshiteru Nakamori (Eds.)
*Creative Environments*, 2007
ISBN 978-3-540-71466-8

Vol. 60. Vladimir G. Ivancevic and Tijana T. Ivacevic
*Computational Mind: A Complex Dynamics Perspective*, 2007
ISBN 978-3-540-71465-1

Vol. 61. Jacques Teller, John R. Lee and Catherine Roussey (Eds.)
*Ontologies for Urban Development*, 2007
ISBN 978-3-540-71975-5

Vol. 62. Lakshmi C. Jain, Raymond A. Tedman and Debra K. Tedman (Eds.)
*Evolution of Teaching and Learning Paradigms in Intelligent Environment*, 2007
ISBN 978-3-540-71973-1

Vol. 63. Wlodzislaw Duch and Jacek Mańdziuk (Eds.)
*Challenges for Computational Intelligence*, 2007
ISBN 978-3-540-71983-0

Vol. 64. Lorenzo Magnani and Ping Li (Eds.)
*Model-Based Reasoning in Science, Technology, and Medicine*, 2007
ISBN 978-3-540-71985-4

Vol. 65. S. Vaidya, Lakhmi C. Jain and Hiro Yoshida (Eds.)
*Advanced Computational Intelligence Paradigms in Healthcare-2*, 2007
ISBN 978-3-540-72374-5

Vol. 66. Lakhmi C. Jain, Vasile Palade and Dipti Srinivasan (Eds.)
*Advances in Evolutionary Computing for System Design*, 2007
ISBN 978-3-540-72376-9

Vol. 67. Vassilis G. Kaburlasos and Gerhard X. Ritter (Eds.)
*Computational Intelligence Based on Lattice Theory*, 2007
ISBN 978-3-540-72686-9

Vol. 68. Cipriano Galindo, Juan-Antonio Fernández-Madrigal and Javier Gonzalez
*Multiple Abstraction Hierarchies for Mobile Robot Operation in Large Environments*, 2007
ISBN 978-3-540-72688-3

Cipriano Galindo
Juan-Antonio Fernández-Madrigal
Javier González

# Multiple Abstraction Hierarchies for Mobile Robot Operation in Large Environments

With 66 Figures

 Springer

Cipriano Galindo
Juan-Antonio Fernández-Madrigal
Javier González
Departamento de Ingeniería de Sistemas y Automática
Universidad de Málaga
ETSI Informática
Campus Teatinos
29071 Málaga (Spain)
*E-mail:-* cipriano@ctima.uma.es
        jafma@ctima.uma.es
        jgonzalez@ctima.uma.es

Library of Congress Control Number: 2007927248

ISSN print edition: 1860-949X
ISSN electronic edition: 1860-9503
ISBN 978-3-540-72688-3 Springer Berlin Heidelberg New York



Springer is a part of Springer Science+Business Media
springer.com
© Springer-Verlag Berlin Heidelberg 2007



Cover design: deblik, Berlin
Typesetting by the SPi using a Springer LaTeX macro package
Printed on acid-free paper     SPIN: 11967842     89/SPi     5 4 3 2 1 0

*To Ana Belén, Ana+, and María José*

# Preface

This book focuses on the efficient performance of mobile robots through the use of a multi-hierarchical and symbolic representation of the environment.

A mobile robot must possess some symbolic representation of its environment when it is intended to carry out deliberative (planned) operations or to reason about the world. However, symbolic representations of real, complex environments usually become so large that it should be conveniently arranged in order to facilitate and, in some cases, make possible its use.

Apart from the drawback of dealing with a large amount of symbolic information, some other problems (not completely solved yet in the scientific literature) stand out when using symbolic representations. One is keeping the symbolic model properly arranged (optimized) with respect to the robot tasks (different models can be constructed for representing a given environment, but not all of them are equally suited for efficient operation). Another is maintaining the model coherent with reality, that is, the creation and maintenance of symbolic data from sensorial data (the so-called symbol grounding problem).

This book addresses these problems through multi-hierarchical structures. Our symbolic structures, based on the concept of *abstraction,* mimic the way in which humans arrange information, allowing a mobile robot to speed up its operation in large environments. Moreover, this book provides practical solutions tested on real robots like a robotic wheelchair for physically impaired or elderly people. For that, a new robotic architecture which includes multi-hierarchical symbolic structures is proposed that permits a high degree of human integration into the robotic system. It enables the wheelchair driver to cooperate with the robot as though both of them were part of the same system.

Málaga,<br>
2007

*Cipriano Galindo*<br>
*Juan-Antonio Fernández-Madrigal*<br>
*Javier González*

# Contents

# List of Figures

# A Robotic Future

> *You see things; and you say 'Why?'*
> *But I dream things that never were; and I say 'Why not?*
> George Bernard Shaw, Irish writer (1856–1950)

*Málaga city*
*Year 2007*
*9:00 am*

*Mark takes a taxi to head for his work place. The robotic driver unit asks him the destination: the high-tech laboratory, he answers. During his daily journey, that usually takes a couple of minutes, he can not help being astonished at what his eyes can see: robots sweeping the street, cleaning up windows, controlling the traffic... Suddenly, RX-342A, the taxi driver, turns left taking an unusual way to the destination. Before Mark could say anything, RX-342A tells him that he has been informed about a traffic problem in the next junction, and thus, he has decided to change the usual route. Mark agrees. Finally, they arrive at the high-tech laboratory on time.*

This vignette represents how we could imagine the future when fiction movies made us believe that the $21^{st}$ century would be full of robots working and living with us (see Fig. 1.1). Today, looking around us, we can discover that great scientific advances have improved our lives, but they are still beyond the described scenario. Thus, we can wonder: Will such an idealized future be, some day, the present? Could engineers reproduce, or at least imitate, human intelligence?

Cipriano Galindo et al.: *A Multi-Hierarchical Symbolic Model of the Environment for Improving Mobile Robot Operation*, Studies in Computational Intelligence (SCI) **68**, 1–8 (2007)
www.springerlink.com                              © Springer-Verlag Berlin Heidelberg 2007

**Fig. 1.1.** Stills from the science fiction movie "Total Recall"

## 1.1 Why is it Not Real?

What prevents machines from working and interacting intelligently with humans? This is the first question we have to consider in order to approach *the robotic future* described in the vignette. From some philosophical points of view, it is stated that the main ingredient that turns humans into intelligent beings is our ability to construct and use symbolic (or *categorical*) representations of our environment [32]. Such representations enable us to carry out intelligent abilities like, for example, to imagine what would be the results of our actions before we perform them. That is, we can simulate and test different strategies to figure out which is the best one according to the current environmental conditions. Moreover, a mental representation of the physical environment simplifies to a great extent the vast amount of information gathered by our senses: our brain is capable of abstracting sensorial information, i.e. the image of a tree, into a general concept, a symbol that represents all trees, regardless of their particular shapes, colors, sizes, etc. This ability allows us to approach efficiently high complex problems without considering the immense amount of information arising from our physical environment. Also, we communicate to each other by sharing symbols.

It seems that in nature there are beings, like insects, that do not account for a symbolic model of their environments: insect behavior seems merely reactive and instinct-based [105] (but it is sufficient for their successful performance). They can automatically perform complex tasks without planning: for example, a spider can spin a spiderweb without using geometric reasoning.

Years ago, the robotics community noticed this simple but effective behavior of insects and tried to imitate it by devising the *reactive framework* [21] (see Fig. 1.2). This framework appeared in the late 1980s as a promising step aimed to endow a robot with a certain degree of intelligence, or at least, with a certain degree of autonomous performance. Reactive systems respond quickly and effectively to external stimuli through wired behaviors that imitate animal instincts.

Reactive systems, as a sum of a number of behaviors, exhibit features that make them suitable to cope with highly dynamic, complex and unpredictable environments. On the one hand, its quick response to external stimuli is remarkable. Wired behaviors can be seen as a pair "perceptual

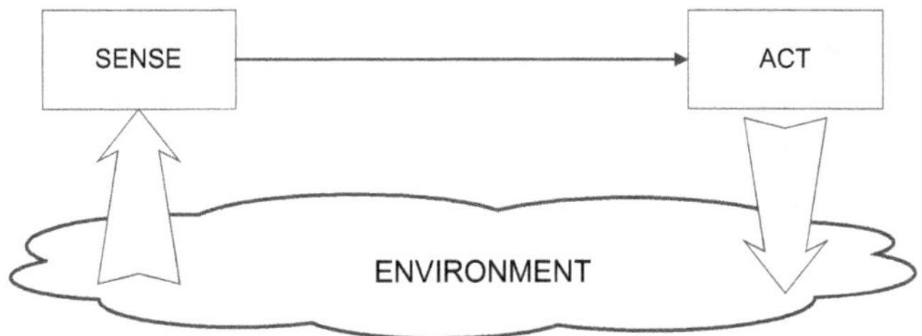


**Fig. 1.2.** The Reactive framework

condition-motor action": once the sensory system yields a perceptual condition, the respective motor action is immediately triggered. Such a quick response is possible due to the processing of only local information. On the other hand, a reactive system treats behaviors as independent entities, which cooperate to produce the overall robot response to its environment. This feature makes reactive systems modular, thus facilitating the combination of a number of robotic behaviors ranging from simple and critical ones like *collision avoidance* to more complex ones like *navigation*, improving the robot's adaptation to its environment. Examples of reactive robotic systems are largely referred to in the literature, for example [6, 73, 104].

Under the reactive framework, in which there is a tight connection between perception and action, the robot does not need a symbolic model of its environment. The main reasons to support avoiding any symbolic knowledge is that no model can accurately represent real and dynamic environments, mainly due to their complex and uncertain nature. Moreover, the construction and maintenance of any approximation is a high time-consuming process. These drawbacks can be categorically summarized through the famous Brook's quote:

*The world is the best representation of itself* [22].

At a first glance, reactive systems may seem the solution to produce a moderately intelligent robot, but can a pure reactive machine perform in the way like our robotic taxi driver? We believe that it certainly cannot.

Analyzing the initial vignette we realize that our robotic driver exhibits certain capabilities that can not be produced under the reactive framework:

- RX-342A has knowledge about streets and junctions, that is, he manages global information to plan the best path to arrive at the destination.
- He can foresee future situations and weigh up his decisions in order to accomplish correctly and efficiently his tasks.
- He communicates his decisions to a human using understandable terms by her/him.

What we can infer from this analysis is that RX-342A must acquire and manipulate symbols that represent physical objects and places from its environment. He recognizes the *high-tech laboratory* as a valid destination and

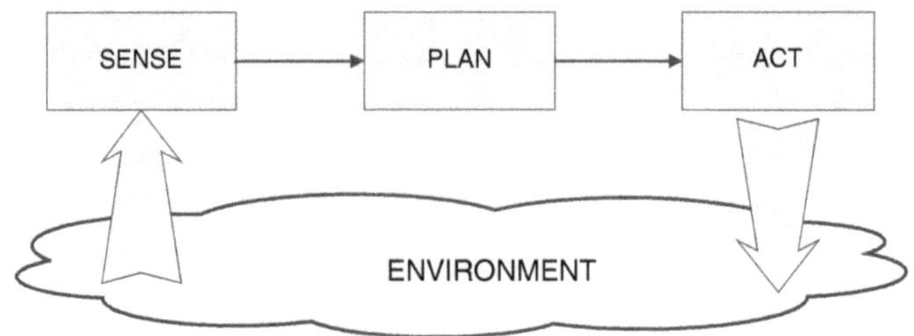


**Fig. 1.3.** The Deliberative framework

plans a path to head for it from the starting point by using other intermediate symbols, like streets and junctions. This is only possible if he accounts for a symbolic internal representation of the whole environment.

Systems that rely on a world model are commonly called *deliberative* systems [112] (see Fig. 1.3). In fact, the *Deliberative Paradigm* was the first attempt to design intelligent robots (even before reactive systems) in the 70's and 80's. Based on the idea that humans, unlike instinct-based animals, plan all our actions, the deliberative paradigm proposes a robotic framework entailing an internal representation (model) of the environment that is used for planning. The intelligent behavior of the robot is achieved then by a planner algorithm, for example *STRIPS* [51], that considers the robot's goal and a global vision of the current state of the environment (through an abstract representation of it) to produce a sequence of actions.

But, is a pure deliberative system just enough? Certainly not again.

The deliberative paradigm is based on the idea that humans plan all our actions. Nowadays it is widely known that we do not plan all what we do, but instead we rely on certain schemes or behaviors to perform some actions. Moreover, pure deliberative systems can only be used when serious assumptions are considered:

- The environment can be accurately modelled.
- Uncertainty is restricted.
- The world remains unchanged during the planning and execution cycle.

It is to be noted that these three assumptions can not be completely ensured in a real and uncontrolled environment, as the one described in the taxi example. In particular, there are approaches to model uncertainty, but they are approximations under certain assumptions (markovian environments, considering not very large scenarios, etc.)

A possible and well accepted solution within the robotic community to this issue is to follow a famous quote from Aristotle:

*In medio stat virtus*[1]

---

[1] All virtue is summed up in dealing justly.

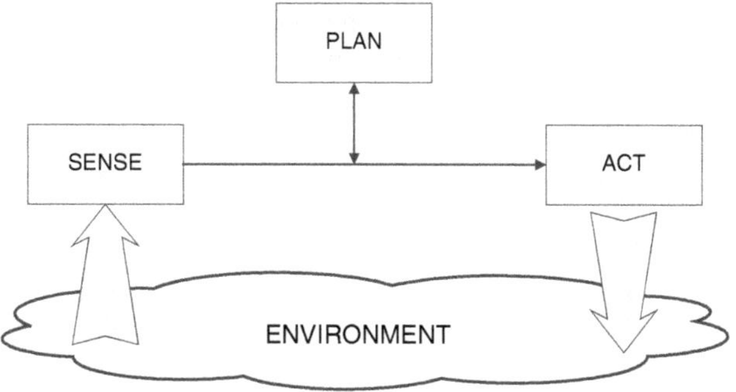



**Fig. 1.4.** Hybrid framework

That is, the best of each system can be adopted to solve our problem. The resultant framework, called the *Hybrid Paradigm* (see Fig. 1.4) has now been widely accepted within the robotic community [7, 137]. Hybrid systems combine aspects from traditional AI, such as symbolic planning and abstract representations, with reactive capabilities to perform robustly under uncertainty. The use of a symbolic model of the robot environment enables it to consider global information to predict the best solution when performing tasks, as well as to adapt its overall behavior to perceived changes. The local use of reactive techniques, for its part, provides robustness and quick operation to the robot, permitting it to react against abnormal and unforeseeable situations, i.e., a child crossing the street, running after her/his ball.

It can be concluded that hybrid frameworks provide a feasible solution to the problem of imitating human intelligence. But, in spite of such a conviction, our initial vignette is still fiction, why? The answer to this question involves a variety of scientific open issues, many of them are out of the scope of this book. However, in our opinion, one of the main reasons for why *the robotic future* is not the present yet is that there is not enough knowledge about the way in which humans construct and use symbolic abstract models, and thus, we can not reproduce this ability in machines. In general, it is clear that the ability for creating symbolic models is learnt by our brain in the first years of our childhood, and that this ability is improved through our life [34, 35]. But neither the internal biological mechanism to acquire new symbols, nor the way in which they are endowed with a particular meaning, are well understood. Then, will *the robotic future* be postponed up to the moment in which the scientific community figures out our most intimate mental processes?

## 1.2 Making it Possible

This book aims to contribute to the design of autonomous and intelligent robots able to work closely with humans. Probably, many years will pass (in the case that it happens someday) before researchers can discover the mental processes involved in our intelligent behavior. Moreover, in the case that this spectacular discovery one day becomes true, it is not really clear the possibility of implementing in computational devices such a mental process to construct intelligent machines [12, 20, 140].

In the meantime, we can approach the problem by taking short but steady steps. Obviously, the reader will not find, at the end of this book, the design of RX-342A, neither thousands of lines of code implementing his intelligent performance, but indeed, she/he could find new ideas, algorithms, mathematical formalizations and code implementations that enable mobile robots to face complex tasks within large environments. Considering this, we will outstandingly simplify the initial vignette turning it into a robotic application in which a mobile robot efficiently plans and carries out tasks, possibly managing volumes of information stemmed from a large indoor scenario, whereas it interacts intelligently with humans. The main elements involved in this approach are:

- *A Mobile robot.* We focus on a mobile robot capable of performing with a certain degree of autonomy within human (and structured) environments, like office buildings, hospitals, etc. The robot has to account for a perceptual system that enables it to extract symbolic information from sensorial data, as well as to detect external stimuli, like obstacles.
- *Hybrid architecture.* The hybrid paradigm is identified as the most convenient framework to design mobile robots meant to perform autonomously and intelligently.
- *Large indoor environments.* The robotic taxi driver (as well as any of us) has to deal with a vast amount of information. All this information comes from experience and should be stored and treated efficiently. In this book we consider large indoor environments, like office buildings or relatively small scenarios in which the robot manages large amounts of information.
- *Planning Efficiency.* Efficiency in planning may turn into one of the bottleneck of hybrid architectures when the robot is supposed to perform in large-scale (or complex) environments. Thus, we strive to reduce as much as possible the computational cost of the planning process.
- *Human Interaction.* As in the taxi vignette, an intelligent robot needs to interact with humans. Such an interaction can be achieved by means of several mechanisms, for example via voice, through tactile or visual information, etc. In any case, the robot should be able to communicate to people using humans concepts, i.e. "please take me to the *High Tech Laboratory*"

The cornerstone of a deliberative agent is the way in which it models the environment. The performance of our robot, and therefore its level of autonomy and intelligence, will largely depend on how it learns, organizes, and uses the symbolic information stemmed from the environment.

But, in spite of the marked benefits that the use of a symbolic representation may provide, it brings some problems among which the following stand out: (i) how the large amount of information arising from a real environment can be managed efficiently, and (ii) how the symbolic representation can be created and maintained along the robot's operational life.

Problem (i) appears in complex scenarios, i.e. large environments like an office building, whose spatial model can contain thousands of concepts and relations between them. In this situation, the robot should account for appropriate mechanisms to efficiently cope with such amount of information. The solution adopted here is to endow the robot with the ability of hierarchically arranging the environmental information, classifying it at different levels of detail.

Problem (ii), the creation and maintenance of the internal model, has not a complete solution yet, since the human mechanism for the creation of symbols that represent physical objects is unknown. This problem is derived from the widely known *symbol grounding problem* [69] in the Psychology literature. The main concern here is how to idealize a symbolic representation from a physical entity, and how to maintain such a model coherent with the world along the robot operation. This issue has been recently approached in the robotic field through *anchoring* [28], which is the solution adopted in this book.

## 1.3 Outline

The outline of this book is as follows.

Chapter 2 describes a mathematical model (based on hierarchies of abstractions) to symbolically represent the robot environment. This model, called *Multi-AH-graph*, has been previously presented in [49], and applied to the robotic field in [47, 56, 58, 59]. In this chapter, a more elegant formulation of the Multi-AH-graph model based on *Category Theory* [127] is given.

Chapter 3 addresses the planning efficiency problem when dealing with large amounts of information. It details two general-purpose task planning approaches that improve outstandingly the robot task planning efficiency when using a hierarchical symbolic model. The proposed planning techniques are compared to other non-hierarchical planning approaches as well as to a well-known hierarchical planner in the robotic field (ABSTRIPS [128]).

Chapter 4 studies the benefits of using multiple hierarchies to arrange symbolic information in different manners, each of them aimed to improve a particular robot operation, i.e. task-planning, self-localization, or human-robot communication [47]. This chapter also explains a symbol translation process, that is, a process that transforms symbols from a certain hierarchy into a symbol (or a set of them) of another hierarchy. This is especially significant when translating symbols from/to the hierarchy used for human-robot communication (the cognitive hierarchy).

Chapter 5 describes a general framework, called *ELVIRA*, which provides a solution to the problem of creating and arranging symbolic information. We describe how a symbolic model can be created automatically or with human assistance, from spatial entities, like rooms, corridors, and simple objects. We also discuss how a possible large amount of symbols can be automatically arranged in order to enhance robot operations, i.e. robot task-planning. Finally, we also consider how the symbolic model can be adapted to reflect changes in the environmental information as well as changes in the robot operational needs, while preserving efficiency in the information processing.

Chapter 6 deals with the design of a hybrid robotic architecture, called *ACHRIN*, that entails the proposed hierarchical and symbolic world model (a Multi-AH-graph) as the principal part of its deliberative tier. ACHRIN has been specifically designed for assistant robots in which there is a clear necessity for providing proper human-robot communication mechanisms. This is achieved by exploiting the high-level communication characteristic that the symbolic model offers. This chapter also shows some experiences carried out on a real mobile robot: a robotic wheelchair called *SENA* [64].

This book is completed by three appendices:

Appendix A deals with mathematical demonstrations regarding the formalization of the categories presented in Chap. 2.

Appendix B presents mathematical demonstrations related to the formalization of the proposed hierarchical task-planning approach under Category Theory presented in Chap. 3.

Finally, Appendix C lists the planning domain considered in the experiences commented along the text.

# Multi-Hierarchical, Symbolic Representation of the Environment

> *Do not worry about your problems with mathematics,*
> *I assure you mine are far greater.*
> Albert Einstein

How much information is stored in our brain? If we could measure it in a certain magnitude, let's say megabytes, how many should be necessary to represent all our knowledge? Moreover, we do not only store a vast amount of data, but we also retrieve and use it efficiently, for instance, when thinking, remembering or solving problems.

It is well stated in the psychologist literature that the key of such intellectual abilities relies on the way in which the human brain arranges the information [68, 71]. If our aim is to develop autonomous robots capable of managing the information arising from a human environment (like an office building), firstly, we have to devise an efficient arrangement of that information.

We have identified that a particular arrangement of information improves operation to a great extent: a hierarchical organization of information based on *abstraction*. The term abstraction can be interpreted in many different ways, but here, abstraction is understood as a process that reduces (abstracts) information through different levels of detail; for instance, in a company, an employee can take part into a work-group that belongs to a certain department, which in its turn reports to a company division, that joined to the rest of divisions makes up the whole company. That is, in this example we consider the following sequence of abstractions: employee $\rightarrow$ work-group $\rightarrow$ department $\rightarrow$ company division $\rightarrow$ company.

In this chapter we describe a mathematical model, called *Multi-AH-graph*, that relies on *abstraction* in the commented sense for modeling knowledge and represents it as annotated graphs [49]. The Multi-AH-graph model, which has been previously presented using classical set theory in [49], is formalized here through *Category Theory*.

Cipriano Galindo et al.: *A Multi-Hierarchical Symbolic Model of the Environment for Improving Mobile Robot Operation*, Studies in Computational Intelligence (SCI) **68**, 9–28 (2007)
www.springerlink.com 

## 2.1 Introduction

Mental processes such as to imagine, to predict, to plan, etc., require a symbolic representation of the involved physical domain. Such a mental view should entail the whole knowledge that the agent, for example a person, has about the physical world. Clearly, the larger the amount of considered information is, the higher the complexity of these processes.

Focusing on robotics and considering, for example, the task-planning capability of the robot as a "mental process", the question is: how could we extend the robot physical domain, that is, its workspace, while keeping a reduced planning complexity?

We can look at our own nature again to devise a solution. Our mental processes, i.e. planning for a trip, manage an immense amount of information, but we usually find quickly a solution, even a variety of possibilities to achieve our goal. It is clear that we, humans, possess some mechanism (an unknown mental process up to now) to store, catalog, treat, and recover all this knowledge. We believe that such a mental process could be (or at least can be imitated through) the so-called *abstraction* [68, 71, 89, 90].

Abstraction is meant here as a mechanism that reduces the amount of information considered for coping with a complex, high-detailed, symbolic representation: symbols that represent a physical entity can be grouped into more general ones and these can be considered as new symbols that can be abstracted again. The result is a *hierarchy of abstraction* or a hierarchy of symbols that ends when all information is modelled by a single universal symbol.

Considering an office building scenario (see Fig. 2.1), it is clear that the amount of information contained in (c) is higher than the information in (b), which in its turn is higher than the information in (a). The information of this scenario can be organized into this hierarchical fashion, and thus we can symbolically represent the entire building with a single symbol, *building*, which comes from the abstraction of a number of symbols that represent each *floor* (like (b)), which in their turn come from the abstraction of a group of symbols representing *corridors, rooms, doors, chairs, cabinets*, ... (like (a)). That is, we can group information at different levels of detail, and thus, we could be able to select the most suitable one for each particular operation.

Let us consider another example to best elucidate how we use abstractions in our daily life. We are planning a trip to visit to our kin at a nearby city (let's call it *city A*). Focusing on our spatial knowledge we can think of the abstract concept "*city A*" which embraces everything about that place. At this point we do not need to consider further information. Next, we take the car and start driving. Now the abstract symbol that represent the city is not enough for us and we need extra information. Thus, we consider more detailed information included into the concept "*city A*", for example roads and city districts. When we approach the area where our kin live we focus on streets and directions to arrive at a particular address: our destination. Finally,

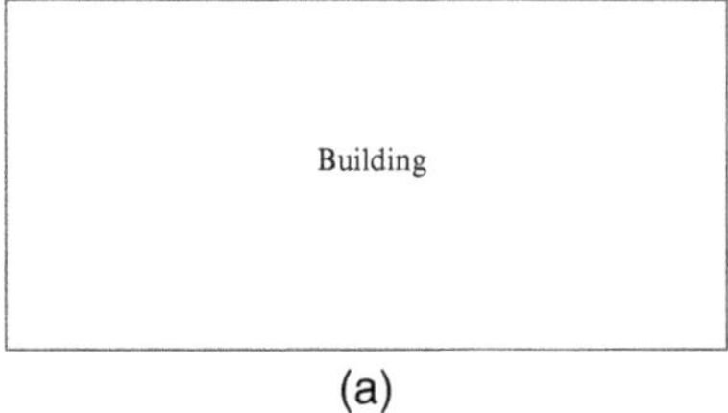



(a)

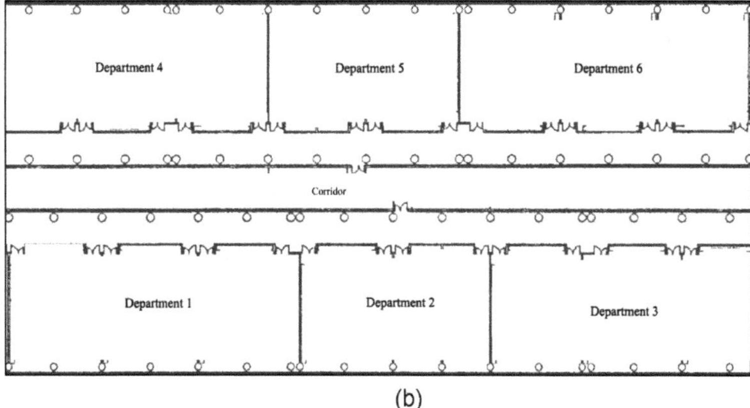



(b)

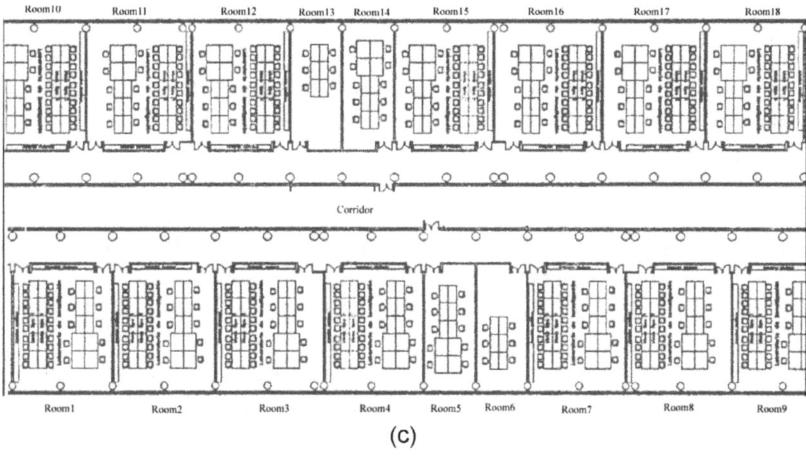



(c)

**Fig. 2.1.** An office environment at different levels of detail: (a) The office building; (b) A particular floor of the building entailing departments. Each department contains a number of rooms connected through corridors; (c) A view of the building with the highest amount of information including rooms and furniture

when we are at the entrance of our kin's house, we can forget everything about cities, roads, districts, streets, etc. to only focus on spatial information about rooms and corridors. This example shows how we abstract (or refine) symbolic information, that is, how we arrange information at different levels of detail using it when needed. Through a hierarchical symbolic model (like the one described further on in this chapter), which simulates the human way to arrange information at different levels of detail, we can efficiently manipulate an immense amount of information considering at each moment only which is relevant for our task.

The hierarchical and symbolic model we describe here has been reified through a mathematical model, called *Multi-AH-graph*, which copes with abstraction in the sense of the previous example. This model is an evolution of simpler relational models: (i) graphs, (ii) annotated or conceptual graphs, and (iii) hierarchical graphs.

(i) A *flat*[1] *graph* model represents relational information efficiently [2]. When flat graphs are used to model physical environments, their vertexes usually represent objects or areas of free space, while their edges represent relations among them. Some representative works on space modelling for robots through graph-based representations are [88, 115, 160].

(ii) Apart from relational information, flat graphs can also store quantitative information. In general, edges can hold *weights* or *costs*, typically a numerical value that is useful to measure the strength of the relation between the pair of vertexes. For instance, an edge can hold the navigational cost (in time or in energy consumption) between two vertexes that represent two locations. But through the use of edge costs there is no way to model other procedural or non-structural data that may be useful: for example, information about the best navigation algorithm to be used by the robot. Such a lack of expressiveness can be overcome by *annotated graphs* (*A-graphs*) that permit the attachment of any piece of information (*annotations*) to both edges and vertexes [49]. Annotated graphs are analogous to conceptual graphs [29].

(iii) A hierarchical graph (*H-graph*) improves the simple relational information that flat graphs can model by arranging them at different levels of abstraction. Each level of abstraction (or detail) is isolated from one another except for the abstraction/refinement links that permit us to move to a more abstracted (and less detailed) point of view of the information (*abstraction*) or to a more refined and detailed level of information (*refinement*) [62]. In case the flat graphs that constitute the H-graph are annotated graphs, that is, their vertexes and/or edges hold annotations, the hierarchical graph becomes an *AH-graph*, that is, a *hierarchical and annotated graph*.

The *AH-graph* model is the fundamental pilar to construct a more powerful spatial model called *Multi-AH-graph*. A Multi-AH-graph [49] deals with multiple hierarchies (or AH-graphs), that is, different ways of arranging hierarchically the same information to improve the management of spatial information

---

[1] The term "flat" denotes that it is not a hierarchical graph.

in different situations. For example, from spatial information about the streets of a city, it is clear that we can maintain (at least) two different views depending on our particular role: if we are driving we consider street directions and speed limitations, while when we are pedestrians we take into account other information like the state of the pavement[2].

In this chapter, the Multi-AH-graph model is mathematically formalized through Category Theory [127]. Firstly, a general and informal description of the Multi-AH-graph model is given. Next a mathematical formalization of: (a) the flat graph model, (b) the category of hierarchical graphs with the abstraction operation (*AGraph*), and (c) its dual category of hierarchical graphs with the refinement operation (*RGraph*), are presented. Particular instances of these categories are the AH-graph and the Multi-AH-graph models. A previous more dense formalization of them can be found in [49], using classical set theory.

## 2.2 Informal Description of a Multi-Hierarchical Model Based on Graphs

The mathematical formalization given in Sects. 2.3–2.5 is useful for a deeply understanding of the rest of this book, however the informal description provided here is enough to get a general flavour of it. First, the single-hierarchy model, called *AH-graph*, is described, and then the multi-hierarchical model (*Multi-AH-graph*), which is made of a number of interconnected *AH-graphs*.

### 2.2.1 What is an AH-Graph?

An *AH-graph* is a relational, graph representation of the environment which includes hierarchical information, that is, the possibility of abstracting groups of elements to *super-elements*. This kind of abstraction produces different layers isolated from one another, called *hierarchical levels*, that represent the same environment at different amounts of detail.

Hierarchical levels in an AH-graph are multigraphs[3]. The lowest hierarchical level of the AH-graph is called the *ground level*, and represents the world with the maximum amount of detail available. The highest hierarchical level is called the *universal level*, and it typically represents the robot environment with a single vertex. Figure 2.2 shows an example of an AH-graph with a single type of relation representing "rigidly joined".

Vertexes of each hierarchical level represent elements (or super-elements) of the world while edges represent relations between them with the possibility of holding weights representing the strength of those relations. For example, in mobile robotics, vertexes can represent distinctive places [91], while edges can

---

[2] Information about pavement state, as well as speed limits, could be modelled as edge annotations.

[3] From now we will use the term "graph" to refer to multigraphs [151].

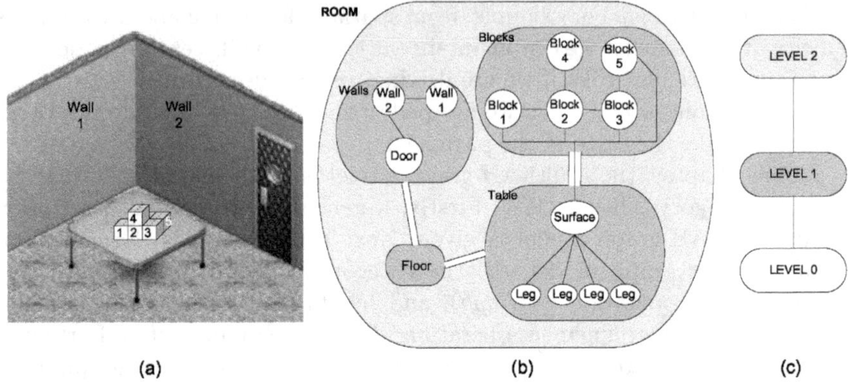

(a)    (b)    (c)

**Fig. 2.2.** An example of a single-hierarchical model (AH-graph) representing a possible abstraction of some spatial elements within a room: (a) A 3D view of a room; (b) Hierarchical levels that model the room (each of them is a flat graph), represented by different grey shades; (c) Resulting hierarchy (each level contains a flat graph)

indicate the navigability relation between them, with the geometric distance as the edge weight.

A group of vertexes of a hierarchical level can be abstracted to a single vertex at the next higher hierarchical level, which becomes their *supervertex* (the original vertexes are called *subvertexes* of that *supervertex*). Analogously, a group of edges of a hierarchical level can be represented by a single edge (their *superedge*) at the next higher level (see Fig. 2.3).

Besides the structural information captured by the AH-graph through vertexes, edges, and hierarchical levels, both vertexes and edges can also hold non-structural information in the form of *annotations* (see Fig. 2.4). This information may include, but is not limited to: geometrical data gathered from the environment (i.e.: maps of obstacles), costs incurred by the robot when executing an action (i.e.: an edge that represents "navigability" from one location to another can store the expected cost energy of that navigation), etc. Non-structural information can be useful for planning and other algorithms. In particular, it is extensively used when the AH-graph model is employed for mobile robot navigation [47].

## 2.2.2 What is a Multi-AH-Graph?

A single hierarchy (AH-graph) is the basis for constructing a multiple hierarchical model upon a common ground level. Broadly speaking, a *Multi-AH-graph* is a set of hierarchies interwoven in a directed acyclic graph structure, where each hierarchy is an AH-graph whose levels can be shared by other hierarchies (see Fig. 2.5). The number of shared hierarchical levels depends upon the power of detecting equivalence between hierarchical levels of the multi-hierarchy [49].

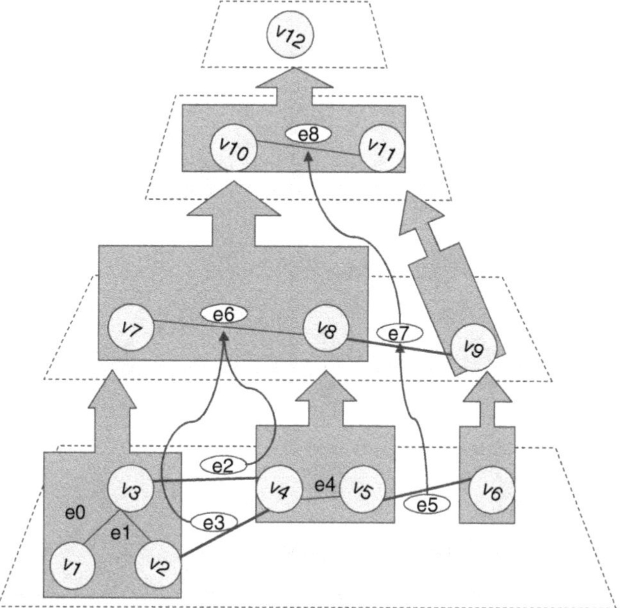


**Fig. 2.3.** An AH-graph example. Vertex abstraction is shown with gray-shadow regions, i.e. vertexes $\{v1,v2,v3\}$ are abstracted into the vertex $v7$. Notice how inner edges (inside each cluster) disappear in the abstraction process, while outer edges (thick lines) are abstracted to edges from the higher levels, i.e., edges $\{e2,e3\}$ are abstracted to the superedge $e6$

Using a symbolic, multi-hierarchical representation of the environment yields three important benefits: first, a multiple hierarchy permits us to choose the best hierarchy to solve each problem (i.e.: to adapt better to diverse problems, improving the overall efficiency, please refer to [50] for more detail); second, when several problems have to be solved, a multiple hierarchy provides the possibility of solving part of them simultaneously; and thirdly, solutions to the problems can be expressed in terms of the concepts of any of the hierarchies, thus the information is given in the most suitable way for each specific purpose, as commented further on. In general, multiple hierarchies have proven to be a more adaptable model than single-hierarchy or non-hierarchical models [49]. This has been recently demonstrated in the particular case of graph search, which has a direct influence on mobile robot route planning [50].

## 2.3 Formalization of Graphs

This section gives a formalization of the classical flat graph model that will serve as the base of our multi-hierarchical model.

A finite, directed, loopless graph $G$ is a tuple:

$$G = (V, E, \gamma, ini, ter)$$

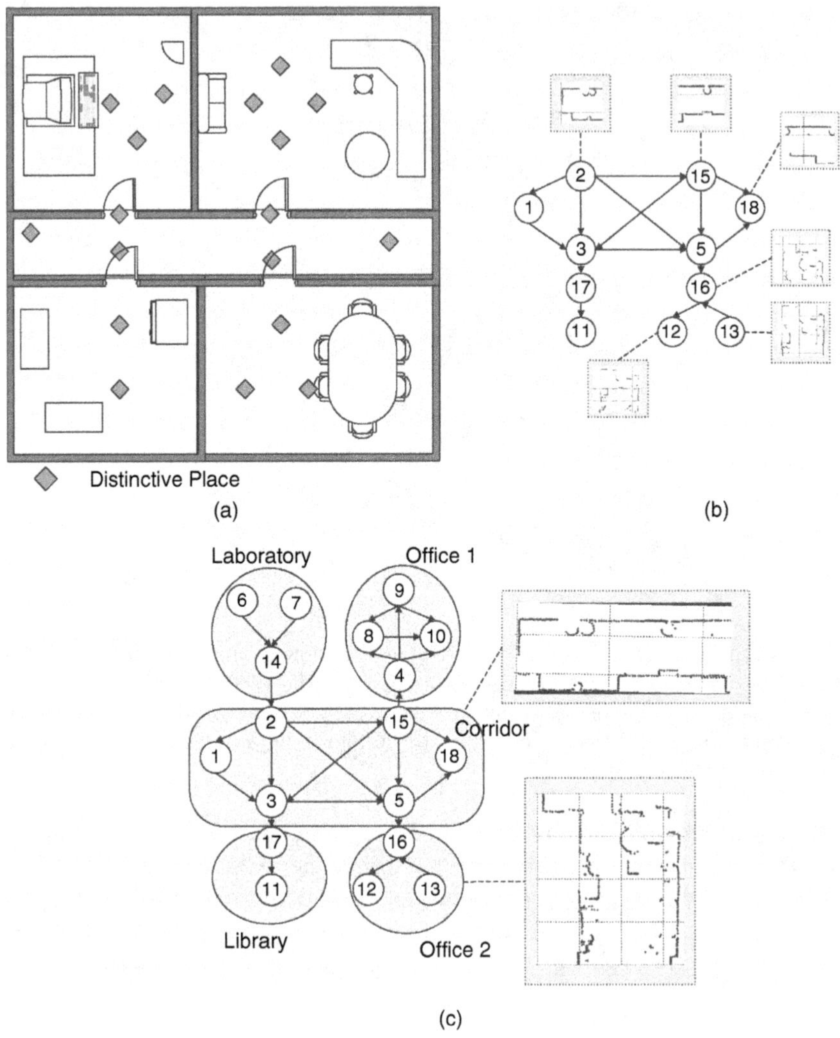



**Fig. 2.4.** Example of an AH-graph holding annotations. This AH-graph models part of a typical office environment (a) through two hierarchical levels. Vertexes at ground level (b) hold point local maps while their supervertexes (c) hold the fusion of a number of them into a global map representing certain areas (rooms)

where $V$ is the finite, non-empty set of *vertexes*, $E$ the finite set of *edges*, $\gamma$ the *incidence function*, *ini* the *initial function*, and *ter* the *terminal function* of the graph. The tuple satisfies the following requirements:

$$V \cap E = \phi, V \neq \phi \tag{2.1}$$

$$\gamma : E \rightarrow (V \times V - \{(a,a) : a \in V\}) \tag{2.2}$$

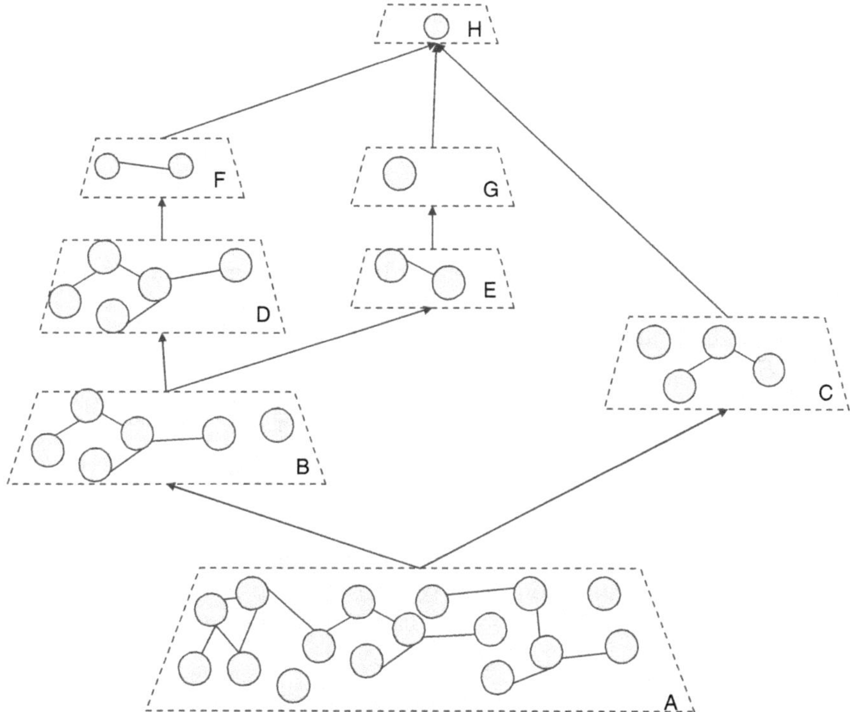


**Fig. 2.5.** An example of multi-hierarchy (with three hierarchies). In this directed acyclic graph, hierarchical levels are represented by dotted rectangular boxes, while arrows represent graph abstractions. Level A is the only ground hierarchical level of this multi-hierarchy, and level H the only universal level

$$ini : E \to V, ter : E \to V, \ such \ that \ \gamma(z) = (ini(z), ter(z)) \qquad (2.3)$$

If the incidence function is on-to[4], the graph is called *complete*. If it is one-to-one[5], the graph is not a graph, but a conventional directed graph, or *digraph* (no parallel edges or loops exists).

The incidence function $\gamma(z)$ yields the vertexes connected by the edge $z$, for instance, $\gamma(z) = (a, b)$. The *initial vertex* of $z$ is the first element of the pair $(a)$, also given by $ini(z)$. The *terminal vertex* of $z$ is the second element of the pair $(b)$, also given by $ter(z)$. Both $a$ and $b$ are said to be *incident* to $z$, and $z$ is said to be *directed from* $a$ to $b$, and to *connect* $a$ and $b$ (or indifferently, to connect $b$ and $a$). Any pair of edges that are directed from $a$ to $b$ are called *parallel*.

---

[4] A function is said to be on-to when every element of the range has a defined correspondence with an element in the domain.

[5] A function is said to be one-to-one when every element of the range that is an image of some element in the domain, is only image of one element.

By constraint 2.2, there may be more than one edge connecting the same pair of vertexes in any direction (parallel edges), but no edge can exist connecting a vertex with itself (*loop*[6]).

By constraint 2.3, functions *ini* and *ter* are directly derived from the definition of the incidence function. Therefore, sometimes they may be omitted for the sake of simplicity. $V$ will be referred to as $V^{(G)}$, $E$ as $E^{(G)}$, and $\gamma$, *ini*, *ter*, as $\gamma^{(G)}$, $ini^{(G)}$, and $ter^{(G)}$ respectively, as long as the graph to which they belong must be specified explicitly.

Vertexes are usually denoted by letters from the beginning of the alphabet: $a, b, c, \ldots$, and edges by single letters from the end: $z, y, z, \ldots$

Two graphs $G$ and $H$ are called *equal*, written $G = H$, iff $V^{(G)} = V^{(H)} \wedge E^{(G)} = E^{(H)} \wedge \gamma^{(G)} = \gamma^{(H)}$ (needless to say: $ini^{(G)} = ini^{(H)} \wedge ter^{(G)} = ter^{(H)}$). Otherwise, they are called *unequal*, written $G \neq H$. Obviously, $G = H \Leftrightarrow H = G, G \neq H \Leftrightarrow H \neq G$.

The set of all the finite, directed graphs without loops will be denoted by $\Theta$.

Finite, directed, loopless graphs will be referred from now on symply as *graphs*. Usually, this type of graphs will be used for representing some portion of knowledge through concepts (vertexes) related by some relationships (edges), thus they are called by some authors *conceptual graphs* [29].


## 2.4 Formalization of Graph Abstraction

An *abstraction from graph G to graph H* is a partial[7] function between both graphs, defined as a tuple:

$$A = (G, H, \nu, \varepsilon)$$

where $G$ is the graph that is abstracted, $H$ is the resulting graph, $\nu$ is the *abstraction function for vertexes*, and $\varepsilon$ is the *abstraction function for edges*. The following restrictions must hold:

$$\nu : V^{(G)} \to V^{(H)} \; is \, a \, partial \, function. \tag{2.4}$$

$$\varepsilon : E^{(G)} \to E^{(H)} \; is \, a \, partial \, function. \tag{2.5}$$

$$\forall z \in E^{(G)}, \; def^8(\varepsilon(z)) \Rightarrow [def(\nu(ini^{(G)}(z))) \wedge def(\nu(ter^{(G)}(z)))] \tag{2.6}$$

---

[6] Those graphs that can contain both parallel edges and loops are called *pseudographs*.

[7] A function is said to be partial when it is not defined for all elements of its domain.

[8] $def(g(x))$ indicates that $g(x)$ is defined, i.e., that element $x$ belongs to the domain of function $g$

$$\forall z \in E^{(G)},\ def(\varepsilon(z)) \Rightarrow [\nu(ini^{(G)}(z)) \neq \nu(ter^{(G)}(z))] \tag{2.7}$$

$$\forall z \in E^{(G)},\ def(\varepsilon(z)) \Rightarrow \begin{bmatrix} \nu(ini^{(G)}(z)) = ini^{(H)}(\varepsilon(z)) \wedge \\ \nu(ter^{(G)}(z)) = ter^{(H)}(\varepsilon(z)) \end{bmatrix} \tag{2.8}$$

Notice that this seems like a conventional graph homomorphism except for the partiality of $\varepsilon$ and $\nu$, and for 2.7.

The vertex $\nu(a)$ for a given vertex $a \in V^{(G)}$ is called the *supervertex* of $a$, or the *abstraction* of $a$. Analogously, the edge $\varepsilon(z)$ for a given edge $z \in E^{(G)}$ is called the *superedge* of $z$, or the abstraction of $z$.

In the case that $\nu$ is total[9], it will be also called *complete*. In the case that $\varepsilon$ is defined for every edge of $G$ except for those whose incident vertexes are abstracted to a same supervertex, it will be called *complete*. If both, $\nu$ and $\varepsilon$ are complete, the whole abstraction is also said to be *complete*. In the case that both $\nu$ and $\varepsilon$ are on-to, the whole abstraction is said to be *covered*. Functions $\nu$ and $\varepsilon$ have inverses defined as:

$$\nu^{-1} : V^{(H)} \to power(V^{(G)})$$
$$\forall b \in V^{(H)},\ \nu^{-1}(b) = \{a \in V^{(G)} : def(\nu(a)) \wedge \nu(a) = b\} \tag{2.9}$$

$$\varepsilon^{-1} : E^{(H)} \to power(E^{(G)})$$
$$\forall y \in E^{(H)},\ \varepsilon^{-1}(y) = \{z \in E^{(G)} : def(\varepsilon(z)) \wedge \varepsilon(z) = y\} \tag{2.10}$$

where we write $power(C)$ to denote the set of all the subsets of $C$. These functions are called *refining functions* for vertexes and edges, respectively.

For any vertex $a \in V^{(H)}$, the vertexes belonging to $v^{-1}(a)$, if any, are called the *subvertexes* of $a$ in $G$. Analogously, for any edge $z \in E^{(H)}$, the edges belonging to $\varepsilon^{-1}(z)$, if any, are called the *subedges* of $z$ in $G$.

Constraints (2.4) and (2.5) define an abstraction as a special kind of morphism between graphs, composed of two partial functions. Notice that it is possible that an abstraction of a graph yields an isomorphic graph or the same graph (for that, both abstraction functions must be total).

Also notice that, by constraints (2.6), an edge can not be abstracted if its incident vertexes are not, and by (2.7) an edge can not either if its vertexes have been abstracted to the same supervertex (in that case, the edge "disappears" in $H$). Constraint (2.8) is the typical definition for *graph homomorphism*: when an edge is abstracted, the incident vertexes of its superedge are the supervertexes of the incident vertexes of the edge (that is, connectivity is preserved).

The collection of all the possible abstractions between any pair of graphs is denoted by $\nabla$. Sometimes we will need to refer to a component of an abstraction specifying explicitly the abstraction to which

---

[9] A function is said to be total when every element of its domain has an image.

it belongs. For that purpose, a superindex will be used. For instance, $G^{(A)}, H^{(A)}, \nu^{(A)}, \varepsilon^{(A)}, [\nu^{-1}]^{(A)}, [\varepsilon^{-1}]^{(A)}$ refer to the components of a particular abstraction $A$.

## 2.5 Category Theory for Abstraction and Refinement of Graphs

Category Theory has itself grown to a branch in mathematics, like algebra and analysis, to facilitate an elegant style of expression and mathematical proofs. A *Category* [38] consists of a collection of objects with a certain structure plus a collection of *arrows* (functions, also called morphisms) that preserve that structure):

$$f : a \rightarrow b$$

Here $a$ and $b$ are objects and $f$ is an arrow (a function) whose source is object $a$ and target is object $b$. Such directional structures occur widely in set theory, algebra, topology, and logic. For example, $a$ and $b$ may be sets and $f$ a total function from $a$ to $b$ or, indeed, $f$ may be a partial function from set $a$ to set $b$; or as we have formalized in this section, objects $a$ and $b$ can be graphs, and $f$ the abstraction of graphs defined in Sect. 2.4.

Among the benefits provided by Category Theory we remark:

- A single result proved in Category Theory generates many results in different areas of mathematics.
- Duality: for every categorical construct, there is a dual, formed by reversing all the morphisms.
- Difficult problems in some areas of mathematics can be translated into (easier) problems in other areas (e.g. by using functors, which map from one category to another)

Next, the category of graphs with abstractions, *AGraph*, as well as its subcategories are formalized. For completeness, the dual of *AGraph*, the category of graphs with refinements (*RGraph*), is also stated.

### 2.5.1 The Category of Graphs with Abstraction

The category of graphs with abstractions, that we will call *AGraph*, is similar to the well-known category *Graph* of digraphs with homomorphisms (very commonly used in graph rewriting and graph grammar literature [29]), except that it is defined under abstractions that are partial morphisms, and it does not allow empty graphs. Some other formulations of the category of graphs under partial morphism exists [79, 109], but they are proposed as a restriction of the existing *Graph* category. Here, we rather propose a direct definition of the new category from scratch.

The *AGraph* category is defined by:

$$AGraph = (\Theta, \triangledown, \ell^-, \ell^+, i, \Diamond)$$

where $\Theta$ is the collection of all possible non-empty, finite, directed graphs without loops, $\triangledown$ is the collection of all possible abstractions on this kind of graphs, $\ell^-$ is the *lower hierarchical level* function, $l^+$ is the *higher hierarchical level* function, i is the *identity* function, and $\Diamond$ is the *composition of abstractions* function, such that:

$$\ell^- : \triangledown \to \Theta$$

is a function that yields the graph that has been abstracted by a given abstraction. That is, $\ell^-((G, H, \nu, \varepsilon)) = G$. Analogously,

$$\ell^+ : \triangledown \to \Theta$$

is a function that yields the graph resulting of a given abstraction. That is, $\ell^+((G, H, \nu, \varepsilon)) = H$.

$$i : \Theta \to \triangledown \tag{2.11}$$

is a function that for any graph G yields an abstraction that leaves it unaltered:

$i(G) = (G, G, \nu_G, \varepsilon_G)$, where[10]:

$$\nu_G : V^{(G)} \to V^{(G)} \qquad \varepsilon_G : E^{(G)} \to E^{(G)}$$

$$\forall a \in V^{(G)}, \nu_G(a) = a \ \forall z \in E^{(G)}, \varepsilon_G(z) = z$$

Finally,

$$\Diamond : \triangledown \times \triangledown \to \triangledown \tag{2.12}$$

is a partial function that yields the composition of two given abstractions $A1$, $A2$ as long as $H^{(A_1)}$ and $G^{(A_2)}$ are equal (otherwise it is undefined). It is constructed as follows:

$$A_2 \Diamond A_1 = (G^{(A_1)}, H^{(A_2)}, \nu_\circ, \varepsilon_\circ)$$

The two abstraction functions are defined by mathematical composition of functions: $\nu_\circ = \nu^{(A_2)} \circ \nu^{(A_1)}$ and $\varepsilon_\circ = \varepsilon^{(A_2)} \circ \varepsilon^{(A_1)}$, that is,

$$\nu_\circ : V^{(G^{(A_1)})} \to V^{(H^{(A_2)})} \text{ (partial)}$$

$$\forall a \in V^{(G^{(A_1)})}, def(\nu_\circ(a)) \Leftrightarrow def(\nu^{(A_1)}(a)) \wedge def(\nu^{(A_2)}(\nu^{(A_1)}(a)))$$

$$\forall a \in V^{(G^{(A_1)})}, def(\nu_\circ(a)) \Rightarrow \nu_\circ(a) = \nu^{(A_2)}(\nu^{(A_1)}(a))$$

$$\varepsilon_\circ : E^{(G^{(A_1)})} \to E^{(H^{(A_2)})} \text{ (partial)}$$

$$\forall z \in E^{(G^{(A_1)})}, def(\varepsilon_\circ(z)) \Leftrightarrow def(\varepsilon^{(A_1)}(z)) \wedge def(\varepsilon^{(A_2)}(\varepsilon^{(A_1)}(z)))$$

$$\forall z \in E^{(G^{(A_1)})}, def(\varepsilon_\circ(z)) \Rightarrow \varepsilon_\circ(z) = \varepsilon^{(A_2)}(\varepsilon^{(A_1)}(z))$$

---

[10] Notice that $\nu_G$ and $\varepsilon_G$ are unique functions for graph $G$.

It can be demonstrated that $A_2 \Diamond A_1$ satisfies constraints for abstractions of graphs (see Appendix A.1).

The composition of abstractions is associative:

$$\forall G, H, J, K \in \Theta,$$
$$\forall A_1 = (G, H, \nu_1, \varepsilon_1), A_2 = (H, J, \nu_2, \varepsilon_2), A_3 = (J, K, \nu_3, \varepsilon_3) \in \nabla, \qquad (2.13)$$
$$(A_3 \Diamond A_2) \Diamond A_1 = A_3 \Diamond (A_2 \Diamond A_1)$$

This can be demonstrated based on the definition of $\Diamond$ (see Appendix A.2).

And finally,

$$\forall G, H \in \Theta, \forall A = (G, H, \nu, \varepsilon) \in \nabla$$
$$A \Diamond \mathrm{i}(G) = A = \mathrm{i}(G) \Diamond A \qquad (2.14)$$

Which can also be demonstrated based on the definitions of $\Diamond$ and i (see Appendix A.3).

Since constraints (2.12), (2.13), and (2.14) are satisfied under this definition of *AGraph*, then *AGraph* is a category.

## Subcategories of AGraph

Depending on certain properties of the abstraction arrow defined on the category AGraph, three different subcategories can be considered (see Fig. 2.6):

*The Subcategory of Complete, Covered Abstractions on Graphs (CVAGraph)*

An interesting subcategory of *AGraph* is *CVAGraph*, also called the *Category of Complete and Covered Abstractions on Graphs*, where every abstraction is complete and covered (Fig. 2.3 in Sect. 2.2 depicts an example of a *CVA-Graph*). The *CVAGraph* is similar to the category Graph widely found in literature (except for on-to abstraction functions, partiality of the abstraction function for edges, and absence of the empty graph).

The first thing to check for considering *CVAGraph* a category is whether it fulfills (2.12), (2.13), and (2.14), or put in a more informal way, to assure

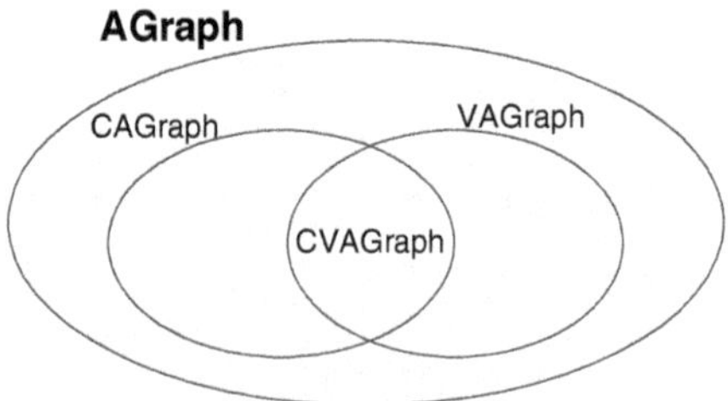


**Fig. 2.6.** Subcategories of the AGraph category. The three subcategories considered here are the subcategory of Complete Abstractions on Graphs (CAGraph), of Covered Abstractions on Graphs (VAGraph), and the subcategory of Complete and Covered Abstractions on Graphs (CVAGraph)

that composition of complete and covered abstractions yields a complete and covered abstraction, that such a composition is associative, and that identity is the neutral element in composition. The latter two issues are satisfied since they hold for any type of abstraction (not only for covered and complete ones) as demonstrated in Appendix A.4.

The first point is easy to see: if $A_1$ and $A_2$ are complete, then so is $A_2 \Diamond A_1$ (if it has sense to define composition), since its domain is the same as $A_1$. On the other hand, if $A_1$ and $A_2$ are covered, then so is $A_2 \Diamond A_1$, since its range is the one of $A_2$.

*The Subcategory of Complete Abstractions on Graphs (CAGraph)*

The *CAGraph* subcategory (see an example in Fig. 2.7) is the same as the *CVAGraph* subcategory except that the abstractions may be covered or not (they must be still complete, as in *CVAGraph*). Under that circumstance, *CAGraph* is a category, since composition of complete abstractions is also a complete abstraction.

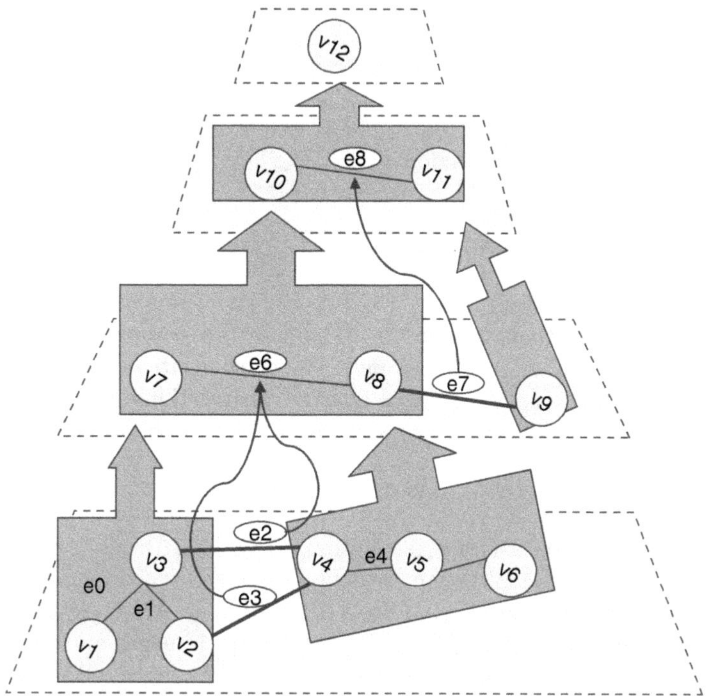


**Fig. 2.7.** A portion of CAGraph. Note that all vertexes and edges of a certain level are abstracted into higher ones. However, neither edge *e7*, nor vertex *v9* have sub-elements, and thus, they are not covered abstractions

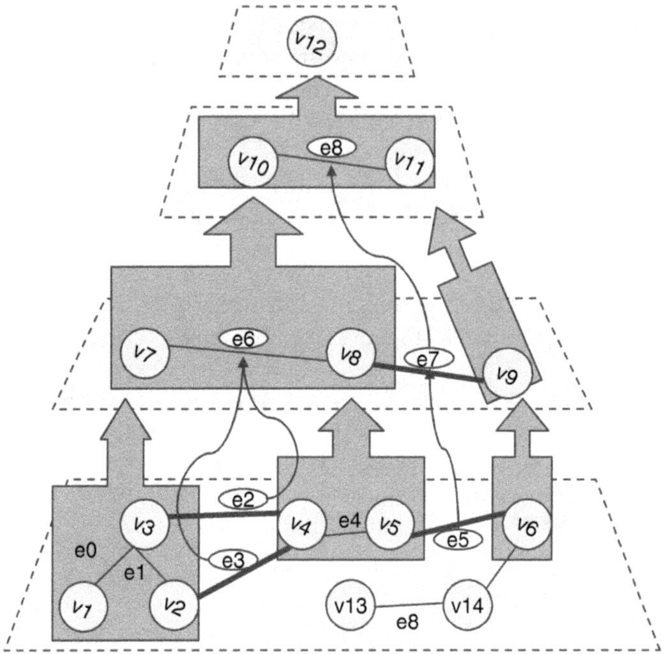

**Fig. 2.8.** A portion of VAGraph. Note that in this case all abstractions are covered, but there are elements (at the ground level) that have not been abstracted, and thus, it is not a complete abstraction

*The Subcategory of Covered Abstractions of Graphs (VAGraph)*

The *VAGraph* subcategory (see Fig. 2.8) is the same as the *CVAgraph* subcategory except that the abstractions may be complete or not (they must be still covered, as in *CVAGraph*). Under that circumstance, *VAGraph* is a category, since composition of covered abstractions is also a covered abstraction.

*AH-Graphs and Multi-AH-Graphs as Portions of AGraph*

Any *Multi-H-Graph* or *H-Graph* is a finite portion of *AGraph*. A partial order relation (total for H-Graphs) can be defined using the abstraction function, that is, $G < H$ iff $\exists A(G, H, \nu, \varepsilon)$. *Multi-AH-Graphs* and *AH-Graphs* are special cases of the previous ones that include annotations in edges and vertexes.

Next, in Sect. 2.5.2, we formalize the category of graphs with refinements (*RGraph*). It can be skipped by the reader due to its resemblance to the *AGraph* category.

### 2.5.2 The Category of Graphs with Refinements

The dual of a category is constructed by reversing every arrow and keeping the same objects and connectivity. The dual of *A Graph* is called the category of graphs with refinements, or *R Graph*. Its definition is based on a new relation between graphs, the *refinement*. A refinement $R$ between two graphs $G$ and $H$ is a tuple:

$$R = (G, H, \mu, \alpha)$$

where $G$ is the graph that is refined, $H$ the resulting graph, $\mu$ is the *refinement function for vertexes*, and $\alpha$ the *refinement function for edges*. The following restrictions must hold:

$$\mu : V^{(G)} \to power(V^{(H)}) \; is \; a \; total \; function. \tag{2.15}$$

$$\alpha : E^{(G)} \to power(E^{(H)}) \; is \; a \; total \; function. \tag{2.16}$$

$$\begin{aligned} \forall a \neq b \in V^{(G)}, \; \mu(a) \cap \mu(b) = \phi \\ \forall y \neq z \in E^{(G)}, \; \alpha(y) \cap \alpha(z) = \phi \end{aligned} \tag{2.17}$$

$$\forall z \in E^{(G)}, \forall y \in \alpha(z), \left[ \begin{array}{l} ini^{(H)}(y) \in \mu(ini^{(G)}(z)) \wedge \\ ter^{(H)}(y) \in \mu(ter^{(G)}(z)) \end{array} \right] \tag{2.18}$$

The collection of every possible refinement between graphs is denoted by $\Delta$. For every abstraction $A = (G, H, \nu, \varepsilon)$ in *A Graph* between two objects $G$ and $H$, there exists one and only one refinement given by $R_A = (H, G, \mu = [\nu^{-1}]^{(A)}, \alpha = [\varepsilon^{-1}]^{(A)})$. Since the definitions for $\nu^{-1}$ and $\varepsilon^{-1}$ given in (2.9) and (2.10) satisfy (2.15–2.18), the category *R Graph* of graphs with refinements is the dual of *A Graph*.

If the abstraction $A$ is covered, its refinement $R_A$ is called *complete*, and it satisfies:

$$\begin{aligned} \forall a \in V^{(H^{(A)})}, \mu^{(R_a)}(a) \neq \phi \\ \forall z \in E^{(H^{(A)})}, \alpha^{(R_a)}(z) \neq \phi \end{aligned} \tag{2.19}$$

And if $A$ is complete, its refinement is called *covered*, and it satisfies:

$$\begin{aligned} \bigcup_{\forall a \in V^{(H^{(A)})}} \mu^{(R_a)}(a) = V^{(G^{(A)})} \\ \bigcup_{\forall z \in E^{(H^{(A)})}} \alpha^{(R_a)}(z) = E^{(G^{(A)})} \end{aligned} \tag{2.20}$$

If $A$ is both covered and complete, then, considering (2.17), (2.19), and (2.20) together, a partition in graph $G^{(A)}$ is induced by $\mu^{(R_a)}$ and $\alpha^{(R_a)}$.

The category *RGraph* can be now defined as a tuple:

$$RGraph = (\Theta, \Delta, \rho^-, \rho^+, i, \bullet)$$

where $\Theta$ is the collection of all possible non-empty, finite, directed graphs without loops, $\Delta$ is the collection of all possible refinements on this kind of graphs, $\rho^-$ is the *abstracted hierarchical level* function, $\rho^+$ is the *refined hierarchical level* function, i is the *identity function*, and $\bullet$ is the *composition of refinements* functions, such that:

$$\rho^- : \Delta \to \Theta \qquad (2.21)$$

is a function that yields the graph that is being refined by a given refinement. That is, $\rho^-((G, H, \mu, \alpha)) = G$. Also,

$$\rho^+ : \Delta \to \Theta$$

is a function that yields the graph resulting of a given refinement. That is, $\rho^+((G, H, \mu, \alpha)) = H$.

$$i : \Theta \to \Delta \qquad (2.22)$$

is a function that for any graph G yields a refinement that leaves it unaltered (or isomorphic):

$$i(G) = (G, G, \mu_G, \alpha_G), \text{ where}^{11}:$$
$$\mu_G : V^{(G)} \to power(V^{(G)}) \quad \alpha_G : E^{(G)} \to power(E^{(G)})$$
$$\forall a \in V^{(G)}, \mu_G(a) = a \qquad \forall z \in E^{(G)}, \alpha_G(z) = z$$

And finally,

$$\bullet : \Delta \times \Delta \to \Delta \qquad (2.23)$$

is a partial function that yields the composition of two given refinements $R_1$, $R_2$ as long as $H^{(R_1)}$ and $G^{(R_2)}$ are equal (otherwise it is undefined). It is constructed as follows:

$$R_2 \bullet R_1 = (G^{(R_1)}, H^{(R_2)}, \mu^{(R_2)} * \mu^{(R_1)}, \alpha^{(R_2)} * \alpha(R_1))$$

where operations $*$ (for edge and vertexes refinement functions) are defined as$^{12}$:

$$\forall a \in V^{(G^{(R_1)})}, \mu^{(R_2)} * \mu^{(R_1)}(a) = \bigcup_{\forall b \in \mu^{(R_1)}(a)} \mu^{(R_2)}(b)$$

$$\forall z \in E^{(G^{(R_1)})}, \alpha^{(R_2)} * \alpha^{(R_1)}(z) = \bigcup_{\forall y \in \alpha^{(R_1)}(z)} \alpha^{(R_2)}(y)$$

---

[11] Notice that $\mu_G$ and $\alpha_G$ are unique functions for graph $G$.

[12] Notice that this special composition of functions can be done only when $H^{(R_1)}$ and $G^{(R_2)}$ are equal.

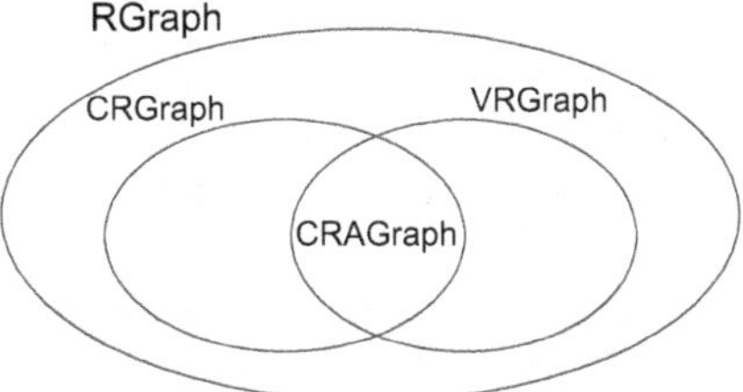


**Fig. 2.9.** Subcategories of the RGraph category

It can be demonstrated that $R_2 \bullet R_1$ satisfies constraints for refinements of graphs (see Appendix A.5).

Composition of refinements is associative:
$\forall G, H, J, K \in \Theta,$

$$\left[ \begin{array}{c} \forall R_1 = (G, H, \mu_1, \alpha_1), R_2 = (H, J, \mu_2, \alpha_2), \\ R_3 = (J, K, \mu_3, \alpha_3) \in \Delta, \\ (R_3 \bullet R_2) \bullet R_1 = R_3 \bullet (R_2 \bullet R_1) \end{array} \right] \tag{2.24}$$

This can be demonstrated based on the definition of $\bullet$ (see Appendix A.6). And finally,

$$\begin{array}{c} \forall G, H \in \Theta, \forall R = (G, H, \mu, \alpha), \\ R \bullet \mathrm{i}(G) = R = \mathrm{i}(G) \bullet R \end{array} \tag{2.25}$$

which can also be demonstrated based on the definitions of $\bullet$ and i (see Appendix A.6.)

Since constraints 2.23, 2.24, and 2.25 are satisfied under this definition of *RGraph*, then *RGraph* is a category. Notice that *RGraph* can also be decomposed in different subcategories (see Fig. 2.9).

These subcategories are the dual of their *AGraph* counterparts: *CVRGraph* is the subcategory of graphs with complete, covered refinements, *CRGraph* is the subcategory of graphs with complete refinements (that is, covered abstractions), and *VRGraph* is the subcategory of graphs with covered refinements (that is, complete abstractions).

## 2.6 Conclusions

This chapter has noticed the relevance of using a certain arrangement of the symbolic information held by an agent for an efficient access and manipulation. This becomes clearly important when the amount of symbolic data is high, which is the case when modeling complex and/or large-scale environments, like an office building.

We have described and mathematically formalized a multi-hierarchical and symbolic world model based on abstractions to arrange spatial, symbolic information. Such a model, called *Multi-AH-graph*, will serve along further chapters as the basis for improving the robot operation within large and human-populated environments.

In the next chapter we use a single hierarchy from the multi-hierarchy to improve efficiency in robot task planning. Such a hierarchical planning approach will be also formalized using *Category Theory*.

# 3

# Application of Hierarchical Models to Mobile Robot Task Planning

*Nothing is particularly hard if you divide it into small jobs.*
Henry Ford

Automatic planning tries to imitate the human decision-making process. In planning, a solution to a problem is given in the form of a sequence of basic actions that transforms a given initial situation of the environment into a desired or goal state.

In the scientific literature, special attention has been paid to issues such as robustness or soundness in planning, but efficiency has usually been pushed to the background (especially in robotics). However, apart from the correctness of solutions, efficiency should also be considered, especially in those applications in which the agent that plans is intended to work with large amounts of information. Let's consider for a moment a variation of the taxi-driver vignette in which the driver must pick up a number of passengers at different places to head for a common destination[1], i.e. an airport. The driver should decide the most convenient path with respect to time, gas consumption, and fares within the city. Under this non-trivial situation, would it be admissible if the driver takes minutes (or hours) for deciding the path to the destination? Although planning is quite simple for human beings, the computational process to plan a task is normally complex, taking large amounts of time and resources (memory). This issue becomes even worse when the planner operates with large environments, making the process very inefficient, even intractable.

In this chapter, we focus on improving the efficiency of the robot task planning process by using our abstraction mechanisms. A general task planning approach which uses abstraction, called *Hierarchical Task Planning through World Abstraction* (HPWA for short) is described. HPWA boosts a general-purpose planner like *Graphplan* [16] or *Metric-FF* [76] by reducing information from the world model that is irrelevant for the tasks to be planned.

---

[1] Note that this is the salesman trip problem which is, like many task-planning problems, NP-complete.

C. Galindo et al.: *A Multi-Hierarchical Symbolic Model of the Environment for Improving Mobile Robot Operation*, Studies in Computational Intelligence (SCI) **68**, 29–63 (2007)
www.springerlink.com

## 3.1 Introduction

Robotic task planning has its origins in the more general problem of automatic planning, which historically constitutes an important issue in Artificial Intelligence, being widely studied since the 1960's. Besides robotics, AI planning has been particularized to a variety of problems such as path and motion planning [84, 93], assembly sequence planning [11, 33], scheduling [138], [161], production planning [142], etc.

In general, the purpose of planning is to synthesize an abstract trajectory in some search space (also named *state space*, since it consists of possible states of the world), predicting outcomes, choosing and organizing actions of different types for reaching goals or for optimizing some utility functions. Maybe surprisingly, there are not many AI classical planners integrated into robotic architectures, mainly due to difficulties in connecting symbolic treatment of information (planning) to non-symbolic sources of data acquired from the real-world. In addition, the uncertainty and partial observability present in this type of applications make classical planning less promising than other approaches [77, 118]. Rather, most of planning in robotics uses specific algorithms intended to guide the execution of very particular robotic tasks (and not others) [4, 117, 141], as is the case of route planning for navigation. Among the few generic (in the AI sense) planners implemented for robotic architectures, some remarkable approaches are STRIPS [51], that was the first planner used for deliberation in a robot (the Shakey robot [21]), and PRODIGY [153], the planner of the Xavier mobile robot.

Few of the works where classical task planners are employed for mobile robots have addressed the problem of computational efficiency, perhaps because they do not deal with a complex and large domain. However, this situation is easily encountered by a real mobile robot that moves within many different places, or when the robot may interact with a lot of world objects, for example, when it is equipped with an on-board manipulator.

Let's think, for instance, of a mobile robot intended to deliver envelopes within an office building (which is the scenario used in our experimental results at the end of this chapter). In this scenario, employees can request the robot from their places to receive their daily post (see Fig. 3.1). Each time a request is commanded to the robot it has to plan a path to arrive to the mail room (maybe opening doors) under time and energy consumption restrictions, search a particular envelope from a rack that may contain hundreds of them (possibly removing others placed on top of the desired one), and finally carry the post to a particular office. Moreover, in a real situation, the robot should also foresee moving to a energy recharger point with a certain frequency in order to keep itself operative.

It seems a simple task that can be easily solved by humans, but it means a tricky goal to be achieved by the planning procedures of a mobile robot. The human advantage to face this type of tasks is our ability (i) to discard irrelevant information with respect to the task, i.e. rooms and envelopes not

**Fig. 3.1.** A typical scenario in which a mobile robot must face tasks that involve a large number of different possibilities within a large-scale environment

involved in the path between the office and the mail room, and (ii) to obtain high level plans which are successively refined, like "go to the mail room", "take the post", or "go to the requested office". A mobile robot lacking from these abilities should explore within a combinatorial space and consider all possible actions up to finding the optimal solution.

In classical planning, a high computational cost arises from this combinatorial search in state space[2]. This cost depends both on the complexity of the description of the states, and on the number of operators that can be applied to each state to obtain another. In the case of robotic task planning, the former corresponds to the symbolic model of the world managed by the robot, while the latter refers to the simple actions that the robot can carry out without planning (for example, move forward, turn, grasp an object, etc.). The lack of efficiency in planning may even lead to the intractability of the problem because of one of these two reasons.

In robotics, task planning efficiency has rarely been addressed. In the AI literature devoted to that issue, the focus has been mostly on managing the possibly large number of operators involved in a complex problem. We believe that the world model deserves more attention in robotics (in pure AI it is usually of a very manageable size), but rather than reducing the number of operators, in real applications like a mobile robot delivering objects in an office building, or a robotic taxi driver working in a large city, the world model should also be reduced.

In this chapter, an approach to improve planning efficiency is presented. It is called *Hierarchical task Planning through World Abstraction* (HPWA for short) [56] and it is particularly oriented towards this kind of large robotic scenarios. The approach consists of breaking down the combinatorial expansion of the search space through the reduction of the size of the description of the world (the model), leaving unchanged the operators. Our method can also yield benefits in other non-robotics areas where planning is carried out on worlds with a large number of elements, such as in Intelligent Transportation Systems.

HPWA takes advantage of the hierarchical arrangement of the world model presented in Chap. 2. In AI literature, abstraction has already been used as a mechanism for improving efficiency of planning. It has been demonstrated that planning with abstraction reduces the search space and it is usually more efficient than non-hierarchical planning [51, 83]. Planners that use abstraction in some way to reduce computational cost are commonly called *hierarchical planners*. A hierarchical planner first solves a problem in a simpler abstract domain and then refines the abstract solution, inserting details that were ignored in the more abstract domain. This is repeated until a final solution is reached. We have found that previous hierarchical planners exploit any of the following three kinds of abstractions [159]: precondition-elimination abstraction, effect abstraction, and task abstraction. Some remarkable implementations are: ABSTRIPS [128], Pablo [27], Prodigy [153] (that uses the hierarchy automatically produced by Alpine [83, 108]), and HTN planners [40] (that use

---

[2] Probability approaches to planning, which constitutes a more modern tendency to task planning under uncertainty, also suffer from efficiency issues, which can also be addressed through abstraction.

task abstraction)[3]. Here, a new kind of hierarchical planning is explored, consisting of abstracting only the description of the world (neither operators nor tasks), which is a different kind of abstraction not used by typical hierarchical planners.

The HPWA approach is basically a procedure that uses an existing planner (the so-called *embedded planner*) at different levels of abstraction of the representation of the world, without changing the internal structure or behavior of such an embedded planner. Although experiments have been run with Graphplan [16] and Metric-FF [76] (they have become very popular planners during the last years), we do not find any limitation in using any other STRIPS-style one. It should be noticed that the embedded planner may already include some other techniques for improving computational efficiency.

Next, Sect. 3.2 gives a formalization of classical planning under Category Theory. The reader not interested in the mathematical details can skip to Sect. 3.3 which presents our first implementation of HPWA. Section 3.4 details a variation of it that permits backtracking when a plan is not found at a certain level. Finally, the reader will find experimental results revealing the improvement achieved with HPWA when compared to other planning approaches in Sect. 3.6.

## 3.2 Formalization of Classical Planning in Category Theory

This section states some definitions needed to formalize our hierarchical planning algorithms for the hierarchical graph model of Chap. 2. For our planning purposes, we use a particular subcategory of the *Complete and Covered Abstractions of Graphs (CVAGraph)* that only considers elements[4] of $\Theta^* \subset \Theta$ whose vertexes are connected at less through an edge. That subset of the *CVAGraph* category is called in the following *CVAGraph**.

In order to formalize our planning approach, firstly we define some basic concepts related to traditional AI planning, like *logical predicate* and *planning state*, which are the base of classical logical planners. Next, we define the operation of *abstraction of planning states*. Based on such definitions, we construct the *Category of Planning States with abstractions*, that we will call *AState*.

Finally, we relate both categories, *CVAgraph** and *AState* through the *Graph-State Functor[5]*. *CVAgraph** is used to hierarchically model the robot environment, while *AState* allows us to plan with that information.

---

[3] Notice that we are concerned here exclusively with non-probabilistic planning.

[4] Remember that $\Theta$ is the collection of all flat graphs, as stated in Sect. 2.3.

[5] A functor between two categories is composed of two functions that map objects and arrows from one category to another.

### 3.2.1 Planning Basics

**Definition 3.2.1 (Logical Predicate)** *A logical predicate is an atomic sentence of a first-order finite language L, with k parameters, that can be defined as a tuple:*

$$p = (\Upsilon, Param)$$

*where $\Upsilon$ is a language string (the* predicate symbol*) that represents a k-ary function, and Param is an ordered list of k language constants which are the parameters of $\Upsilon$. In general, we will also denote as $\Upsilon$ the set of all possible strings that represents predicate symbols in the language, and as Param the set of all possible parameters for such predicates. Thus, for referring to the components of a given predicate p, we will rather write $\Upsilon^{(p)}$ and $Param^{(p)}$.*

*The number of parameters of a predicate p is denoted as $length(Param^{(p)})$, while the i-th parameter of the ordered list $Param^{(p)}$ will be referred as $Param_i^{(p)}$, being $i \in [1 \ldots length(Param^{(p)})]$.*

**Definition 3.2.2 (Planning State)** *A planning state (a state for short) is a finite and consistent[6] set of logical predicates joined through two logical connectives: and ($\wedge$), or ($\vee$), that represents some world information. We denote as $\mho$ the set of all possible planning states over our language L. Informally, planning is the process that transforms an* initial state, *that represents the current information available from the world, into a* goal state, *that represents the desired final situation.*

***Remark 3.2.1*** *Given a planning state $S \in \mho$, we can define the* State Parameters *function, $SP(S)$, as a function that yields a set containing the distinct parameters from all logical predicates of S. More formally:*

$$SP : \mho \rightarrow Power(Param)$$

$$SP(S) = \bigcup_{p \in S} Param^{(p)}$$

*Similarly, we define the* State Predicate Symbols *function, $SN(S)$, that yields a set containing all distinct predicate symbols of S:*

$$SN : \mho \rightarrow Power(\Upsilon)$$

$$SN(S) = \bigcup_{p \in S} \Upsilon^{(p)}$$

---

[6] That is, a set of predicates with a consistent interpretation under a certain domain. Formally, a state $S$ composed of $n$ predicates (or formulas) is *consistent* if there is no formula $p$ such that $S \vdash p$ (that is, $p$ is inferred from $S$) and $S \vdash \neg p$ (*not* $p$ is inferred from $S$). In other words, a state is consistent if it does not entail a pair of contradictory opposite predicates [132].

**Definition 3.2.3 (Operator)** *An operator is a pair $\langle Precond, Postcond \rangle$, where Precond is a set of logical predicates representing the conditions under which the operator is applicable, while $Postcond = (Add, Del)$ contains two sets of logical predicates Add and Del that will be added and removed respectively from the planning state in which the operator is applied for obtaining a resulting state. Thus, given a planning state $S$, an operator $o$ can be applied if and only if the set of the logical predicates of its preconditions are true under $S$ through a given instantiation of the parameters of $Precond^{(o)}$. After the application of the operator, state $S$ is transformed into $S'$, using the set of postconditions $Postcond^{(o)}$ instantiated in the same manner: $S' = S \cup Add^{(o)} - Del^{(o)}$.*

**Definition 3.2.4 (Problem Space, Planning Process, Plan)**
*A problem space is composed of a first-order language $(L)$, a initial state $(S_i)$, a goal state $(S_g)$, and a set of finite operators $(O)$. Within a certain problem space, the* planning process *consists of searching a chained sequence of operators, $o_1, o_2, \ldots, o_n$, that transforms the initial state $S_i$ into the goal state $S_g$. Such a sequence of operators is commonly called a plan.*

### 3.2.2 The Category of Planning States with Abstraction

This section formalizes the category of planning states with abstraction in a similar way to the one presented in Sect. 2.5. A reader not interested in this formalization can skip it and go directly to Sect. 3.3.

**Definition 3.2.5 (Abstraction of Planning States)**
*An abstraction from a planning state $S$ to a planning state $T$ is a morphism between both states, defined as a tuple:*

$$A_s = (S, T, \xi, \pi), where$$

$$\xi : SN(S) \to SN(T) \; is \; a \; partial \; function \tag{3.1}$$

$$\pi : SP(S) \to SP(T) \; is \; a \; partial \; function \tag{3.2}$$

$$\forall p \in S, def(\xi(\Upsilon^{(p)})) \Rightarrow \forall m \in Param^{(p)}, def(\pi(m)) \tag{3.3}$$

$$\forall s \in T, \exists p \in S :: def(\xi(\Upsilon^{(p)})) \land \Upsilon^{(s)} = \xi(\Upsilon^{(p)}) \land$$
$$length(Param^{(p)}) = length(Param^{(s)}) \land \tag{3.4}$$
$$\forall i \in [1..length(Param^{(p)})]\pi(Param_i^{(p)}) = Param_i^{(s)}$$

**Definition 3.2.6 (The Category of Planning States with Abstractions)**
*Similarly to the definition of the category of graphs with abstractions presented in Chap. 2, we formulate here the category of planning states with abstractions,* AState.

*Given a first-order language[7] L:*

$$AState(L) = (\mho, \blacktriangle, \varpi^-, \varpi^+, \mathrm{i}, \Diamond)$$

where $\mho$ is the collection of all possible states defined in $L$, $\blacktriangle$ is the collection of all possible abstractions on those states, $\varpi^-$ is the *refined state* function, $\varpi^+$ is the *abstracted state* function, i is the *identity* function, and $\Diamond$ is the *composition of state abstractions* function, such that:

$$\varpi^- : \blacktriangle \to \mho$$

is a function that yields the planning state that has been abstracted by a given abstraction. That is, $\varpi^-((S, T, \xi, \pi)) = S$. Analogously,

$$\varpi^+ : \blacktriangle \to \mho$$

is a function that yields the planning state resulting of a given abstraction. That is, $\varpi^+((S, T, \xi, \pi)) = T$. Besides,

$$\mathrm{i} : \mho \to \blacktriangle \tag{3.5}$$

is a function that for any state $S$ yields an abstraction that leaves it unaltered:

$$\mathrm{i}(S) = (S, S, \xi_S, \pi_S), \text{ where:}$$
$$\xi_S : SN(S) \to SN(S) \;\; \pi_S : SP(S) \to SP(S)$$
$$\forall p \in SN(S), \xi_S(p) = p \;\; \forall a \in SP(S), \pi_S(a) = a$$

And finally,

$$\Diamond : \blacktriangle \times \blacktriangle \to \blacktriangle \tag{3.6}$$

is a partial function that yields the composition of two given abstractions $A_{s1}$, $A_{s2}$ as long as $T^{(A_{s1})} = S^{(A_{s2})}$. It is constructed as follows:

$$A_{s2}\Diamond A_{s1} = (S^{(A_{s1})}, T^{(A_{s2})}, \xi_\circ, \pi_\circ)$$

The two abstraction functions, $\xi_\circ$ and $\pi_\circ$, are defined by mathematical composition, $\xi_\circ = \xi^{(A_{s2})} \circ \xi^{(A_{s1})}$ and $\pi_\circ = \pi^{(A_{s2})} \circ \pi^{(A_{s1})}$, such that:

$$\xi_\circ : SN(S^{(A_{s1})}) \to SN(T^{(A_{s2})}) \, (partial)$$
$$\forall p \in SN(S^{(A_{s1})}), def(\xi_\circ(p)) \Leftrightarrow def(\xi^{(A_{s1})}(p)) \wedge def(\xi^{(A_{s2})}(\xi^{(A_{s1})}(p)))$$
$$\forall p \in SN(S^{(A_{s1})}), def(\xi_\circ(p)) \Rightarrow \xi_\circ(p) = \xi^{(A_{s2})}(\xi^{(A_{s1})}(p)) \tag{3.7}$$

---

[7] In the rest, we assume that *AState* is defined on a certain first-order language $L$ which will not be explicitly specified any more.

$\pi_\circ : SP(S^{(A_{s1})}) \to SP(T^{(A_{s2})})\ (partial)$

$$\forall m \in SP(S^{(A_{s1})}), def(\pi_\circ(m)) \Rightarrow \pi_\circ(m) \Leftrightarrow def(\pi^{(A_{s2})}(\pi^{(A_{s1})}(m))) \qquad (3.8)$$

$$\forall m \in SP(S^{(A_{s1})}), def(\pi_\circ(m)) \Rightarrow \pi_\circ(m) = \pi^{(A_{s2})}(\pi^{(A_{s1})}(m))$$

The composition of abstractions is associative:

$$\forall P, Q, R, S \in \mho,$$
$$\forall A_{s1} = (P, Q, \xi_1, \pi_1), A_{s2} = (Q, R, \xi_2, \pi_2), A_{s3} = (R, S, \xi_3, \pi_3) \in \blacktriangle, \qquad (3.9)$$
$$(A_{s3}\Diamond A_{s2})\Diamond A_{s1} = A_{s3}\Diamond(A_{s2}\Diamond A_{s1})$$

It also must be that:

$$\forall S \in \mho, \forall A_s = (S, T, \xi, \pi) \in \blacktriangle$$
$$A_s\Diamond i(S) = A_s = i(S)\Diamond A_s \qquad (3.10)$$

Since constraints (3.6), (3.9), and (3.10) are satisfied under this definition of *AState*, then *AState* is a category. Demonstrations of these constraints are similar to those given in Appendix A.1 (please refer to it).

A dual category, the *Category of Planning States with Refinements* (*RState*) can be also formulated by reversing every arrow of *AState*. Its formalization is analogous to the one presented here.

### 3.2.3 Functors Between the *AGraph* and the *AState* Categories

Informally, *functors* are functions that relate objects and arrows between two categories, preserving their structures. In this section we state a relation between objects and abstractions from *AGraph* (more precisely from *CV AGraph**) and *AState* in such a way that a graph will correspond to a state. This relation will serve to formalize our hierarchical planning approaches.

But, previous to the formalization of the functor that relates *CV AGraph** to *AState*, we have to provide some auxiliary definitions. Definitions (3.2.7) and (3.2.8) provide the *CV AGraph** category with a system of types for both edges and vertexes. Definition (3.2.9) takes these types as a medium of linking graphs to states. Finally, definition (3.2.10) constructs the complete functor from *CV AGraph** to *AState*.

**Definition 3.2.7 (Edge-Predicate Translator)** *The* Edge-Predicate Translator *function,* $\Gamma_e$, *enriches the CV AGraph* category, being the first step to transform graph edges into predicates. It is a partial function from the set of all possible set of edges of graphs[8] in CV Agraph*, power($E^{(\Theta^*)}$), to triples of all the possible predicate symbols* $\Upsilon$ *involved in* AState. *That is:*

---

[8] For the sake of simplicity, we denote with $E^{(\Theta^*)}$ the set of all those edges. More formally: $E^{(\Theta^*)} = \bigcup_{G \in \Theta^*} E^{(G)}$.

$$\Gamma_e : power(E^{(\Theta^*)}) \to \Upsilon \times \Upsilon \times \Upsilon$$

$$\forall Z \in power(E^{(\Theta^*)}), def(\Gamma_e(Z)) \Rightarrow \Gamma_e(Z) = (g,\ h,\ i) : g,\ h,\ i \in \Upsilon$$

*We impose that triples yielded by $\Gamma_e$ must cover separatively the range of the function, that is:*

$$\left[ \bigcup_{\{Z_i \in power(E^{(\Theta^*)}):def(\Gamma_e(Z_i)),\Gamma_e(Z_i)=(g,h,i)\}} g \right] = \Upsilon$$

$$\left[ \bigcup_{\{Z_i \in power(E^{(\Theta^*)}):def(\Gamma_e(Z_i)),\Gamma_e(Z_i)=(g,h,i)\}} h \right] = \Upsilon \qquad (3.11)$$

$$\left[ \bigcup_{\{Z_i \in power(E^{(\Theta^*)}):def(\Gamma_e(Z_i)),\Gamma_e(Z_i)=(g,h,i)\}} i \right] = \Upsilon$$

*Notice that through $\Gamma_e$ we can define a triple of predicate symbols for a set of edges taken from different graphs. A set of edges will correspond to three predicate symbols, the first and second one representing a type for their initial and terminal vertexes, respectively, while the third one represents a type for any edge of the set. In order to cover all the edges of $E^{(\Theta^*)}$ with $\Gamma_e$, we impose it to define a partition over $E^{(\Theta^*)}$, and thus, the following restrictions must hold:*

$$\left[ \bigcup_{\{Z_i \in power(E^{(\Theta^*)}):def(\Gamma_e(Z_i))\}} Z_i \right] = E^{(\Theta^*)}$$

$$\left[ \bigcap_{\{Z_i \in power(E^{(\Theta^*)}):def(\Gamma_e(Z_i))\}} Z_i \right] = \emptyset$$

$$\forall Z \in power(E^{(\Theta^*)}), def(\Gamma_e(Z)) \Rightarrow Z \neq \emptyset$$

*We also define the following three total functions for retrieving separately each of the three predicates yielded by $\Gamma_e$ (they are total since $\Gamma_e$ induces a partition in $E^{(\Theta^*)}$, and also onto by 3.11):*

$$\Gamma_{e1} : E^{(\Theta^*)} \to \Upsilon$$
$$\forall z \in E^{(\Theta^*)}, \Gamma_{e1}(z) = g : \exists Z \in power(E^{(\Theta^*)}) \wedge z \in Z \wedge$$
$$def(\Gamma_e(Z)) \wedge \Gamma_e(Z) = (g,\ h,\ i)$$

$$\Gamma_{e2} : E^{(\Theta^*)} \to \Upsilon$$
$$\forall z \in E^{(\Theta^*)}, \Gamma_{e2}(z) = h : \exists Z \in power(E^{(\Theta^*)}) \wedge z \in Z \wedge$$
$$def(\Gamma_e(Z)) \wedge \Gamma_e(Z) = (g,\ h,\ i)$$

$$\Gamma_{e3} : E^{(\Theta^*)} \to \Upsilon$$

$$\forall z \in E^{(\Theta^*)}, \Gamma_{e3}(z) = i : \exists Z \in power(E^{(\Theta^*)}) \land z \in Z \land$$

$$def(\Gamma_e(Z)) \land \Gamma_e(Z) = (g,\, h,\, i)$$

*As commented before, $\Gamma_e$ yields a type for edges and for their initial and terminal vertexes. However, sometimes it can be necessary to obtain a predicate from a set of connected edges of different types. For that, we will also need a partial function $\Gamma_e^*$ that yields a language string given any set of edges:*

$$\Gamma_e^* : power(E^{(\Theta^*)}) \to \Upsilon$$

**Remark 3.2.2** *Functions $\Gamma_e$ and $\Gamma_e^*$ are functions that convert sets of edges of graphs into predicate symbols that represent the same graph-based relational information but in terms of a predicate-based language.*

**Remark 3.2.3** *Since $\Gamma_{e1}$, $\Gamma_{e2}$, $\Gamma_{e3}$, and $\Gamma_e^*$ are onto, their inverses are always defined.*

**Definition 3.2.8 (Vertex-Param Translator)** *The* Vertex-Param Translator *function $\Gamma_v$ also enriches the $CVAGraph^*$ category, being a partial, one-to-one, and onto function from the set of vertexes of graphs[9], $V^{(\Theta^*)}$, to the set of all possible parameters $Param$ of predicates of our first order language:*

$$\Gamma_v : V^{(\Theta^*)} \to Param$$

$$\forall a \in V^{(\Theta^*)}, \Gamma_v(a) = r\ : r \in Param$$

**Remark 3.2.4** *We will consider in the rest of this chapter that $\Gamma_v$ is defined for all the vertexes of the graphs we will deal with.*

**Remark 3.2.5** *Since $\Gamma_v$ is defined as a onto there trivially exists its inverse, denoted $\Gamma_v^{-1}$, and since it is one-to-one, the inverse always yields one vertex.*

**Definition 3.2.9 (Edge-State Translator)** *The* Edge-State *translator $\beta_G$ is a total function based on the definition of $\Gamma_e$, $\Gamma_e^*$, and $\Gamma_v$, that yields a set of logical predicates that represent the state of the world corresponding to the subgraphs of a given graph $G$. It is defined based on other four auxiliary total functions, $\beta_G^1$, $\beta_G^2$, $\beta_G^3$, and $\beta_G^4$:*

$$\beta_G : power(E^{(G)}) \to \mho$$

$$\forall Z \in power(E^{(G)}),\ \beta_G(Z) = \left[\bigcup_{z \in Z} \left[\beta_G^1(z) \cup \beta_G^2(z) \cup \beta_G^3(z)\right]\right] \cup \beta_G^4(Z),\ \text{where}$$

---

[9] As before, $V^{(\Theta^*)} = \bigcup_{G \in \Theta^*} V^{(G)}$.

$$\beta_G^1 : E^{(G)} \to \mho$$

$$\forall z \in E^{(G)}, \beta_G^1(z) = \{(\Gamma_{e1}(z) \, \Gamma_v(ini(z)))\}$$

$$\beta_G^2 : E^{(G)} \to \mho$$

$$\forall z \in E^{(G)}, \beta_G^2(z) = \{(\Gamma_{e2}(z) \, \Gamma_v(ter(z)))\}$$

$$\beta_G^3 : E^{(G)} \to \mho$$

$$\forall z \in E^{(G)}, \beta_G^3(z) = \{(\Gamma_{e3}(z) \, \Gamma_v(ini(z)) \, \Gamma_v(ter(z)))\}$$

$$\beta_G^4 : power(E^{(G)}) \to \mho$$

$$\forall Z \in power(E^{(G)}), def(\Gamma_e^*(Z)) \Rightarrow \beta_G^4(Z)$$

$$= \{\Gamma_e^*(Z) \, a_1, a_2, \ldots, a_n) : a_i \in \{\Gamma_v(ini(z)) : z \in Z\}$$

$$\cup \{\Gamma_v(ter(z)) : z \in Z\} \wedge \forall i \neq j, a_i \neq a_j, \}$$

$$\forall Z \in power(E^{(G)}), \neg def(\Gamma_e^*(Z)) \Rightarrow \beta_G^4(Z) = \emptyset$$

*Informally, $\beta_G$ relates the types of edges and vertexes of the graph $G$ to predicates: $\beta_G^1$ transforms an edge into a predicate with one parameter (the initial vertex of the edge), $\beta_G^2$ transforms and edge into another unary predicate (the parameter is the terminal vertex), and $\beta_G^3$ transforms an edge into a predicate with two parameters (both vertexes). In its turn, $\beta_G^4$ transforms the subgraph induced by a set of edges into a single predicate with as many parameters as distinct vertexes are in that subgraph, if $\Gamma_e^*$ is defined for that set of edges. This covers most of the possibilities of transforming a graph that represents a portion of the world into a state with equivalent information.*

Table 3.2 and equation 3.12 below show the application of the *Edge-State* translator to some subgraphs of three sample graphs.

In our planning domains (that is, under our definitions of $\Gamma_e$), the planning state yielded by $\beta_G(z)$ (the first case in Fig. 3.2) might be: $\{$`(is-an-object a)`, `(is-a-place b)`, `(at a b)`$\}$, where *is-an-object, is-a-place, at* $\in \Upsilon : \Gamma_{e1}(z) =$*is-an-object*, $\Gamma_{e2}(z) =$*is-a-place*, $\Gamma_{e3}(z) = at$, and $\Gamma_e^*(\{z\}) = at$. That is, in this example, edge $z$ indicates the position of an object in the robot workspace.

Now, with the previous defined functions we can formalize the *Graph-State Functor* for the categories $CVAGraph^*$ and $AState$.

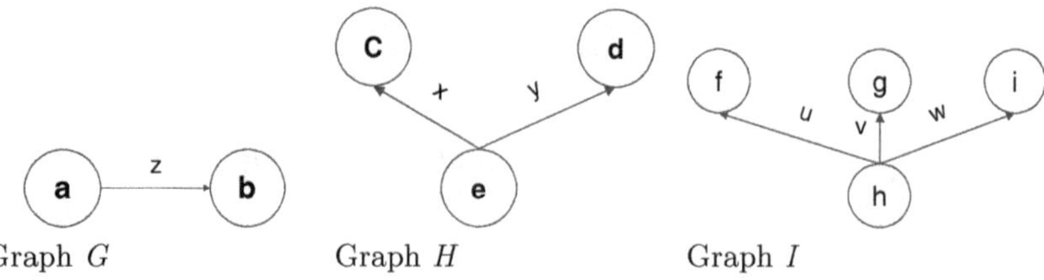


**Fig. 3.2.** Examples for the application of the Edge-State translator for three graphs, that is, $\beta_G$, $\beta_H$, and $\beta_I$.

$$\beta_G(\{z\}) = \{\beta_G^1(z) \cup \beta_G^2(z) \cup \beta_G^3(z) \cup \beta_G^4(\{z\})\}$$
$$= \{(\Gamma_{e1}(z)\, a), (\Gamma_{e2}(z)\, b), (\Gamma_{e3}(z)\, a\, b), (\Gamma_e^*(z)\, a\, b)\}$$
$$\beta_H(\{x, y\}) = \{\beta_H^1(x) \cup \beta_H^2(x) \cup \beta_H^3(x) \cup \beta_H^1(y) \cup \beta_H^2(y) \cup \beta_H^3(y) \cup \beta_H^4(\{x, y\})\}$$
$$= \{(\Gamma_{e1}(x)\, c), (\Gamma_{e2}(x)\, e), (\Gamma_{e3}(x)\, c\, e), (\Gamma_e^*(x)\, c\, e),$$
$$(\Gamma_{e1}(y)\, e), (\Gamma_{e2}(y)\, d), (\Gamma_{e3}(y)\, e\, d), (\Gamma_e^*(\{x, y\})\, c\, d\, e)\}$$
$$\beta_I(\{z\}) = \{\beta_I^1(z) \cup \beta_I^2(z) \cup \beta_I^3(z) \cup \beta_I^4(\{z\})\}$$
$$= \{(\Gamma_{e1}(z)\, d), (\Gamma_{e2}(z)\, a), (\Gamma_{e3}(z)\, d\, a), (\Gamma_e^*(z)\, d\, a)\}$$
$$\beta_I(\{v, w\}) = \{\beta_I^1(v) \cup \beta_I^2(v) \cup \beta_I^3(v) \cup \beta_I^1(w) \cup \beta_I^2(w) \cup \beta_I^3(w) \cup \beta_I^4(\{v, w\})\}$$
$$= \{(\Gamma_{e1}(v)\, h), (\Gamma_{e2}(v)\, g), (\Gamma_{e3}(v)\, h\, g), (\Gamma_{e1}(w)\, h), (\Gamma_{e2}(w)\, i),$$
$$(\Gamma_{e3}(w)\, h\, i), (\Gamma_e^*(\{v, w\})\, g\, h\, i)\} \tag{3.12}$$

**Definition 3.2.10 (Graph-State Functor)** *Functors are maps between categories that preserve their structures. A functor $F$ between two categories, i.e. $A$ and $B$, consists of two functions, one from the objects of category $A$ to those of $B$, and one from the arrows of $A$ to those of $B$. It is usual to denote both of these functions by the functor name. A functor $F$ from category $A$ to $B$ must satisfy the following restrictions:*

$$\forall x \in Obj(A), F(\mathrm{i}(x)) = \mathrm{i}(F(x)), and$$
$$\forall g, f \in Arrow(A), F(f \circ g) = F(f) \circ F(g) \tag{3.13}$$

*where $Obj(A)$ and $Arrow(A)$ denote the set of all objects and the set of all arrows respectively from category $A$.*

*Therefore, we define the* Graph-State Functor, *$\Psi = (\Psi_o, \Psi_a)$, between the categories* CVAGraph* *and* AState, *as follows:*

$$\Psi_o : \Theta^* \to \mho$$
$$\Psi_a : \triangledown \to \blacktriangle$$

*being both of them total functions.*

Informally, $\Psi$ permits us to transform a graph into a state (a set of logical predicates) and an abstraction of graphs into an abstraction of states that preserves the former transformation.

Using the *Edge-State* translator $\beta_G$, we can formalize the *Graph-State Functor* for objects from *CVAGraph** and *AState* (noted as $\Psi_o$) as follows:

$$\Psi_o : \Theta^* \to \mho$$
$$\forall G \in \Theta^*, \Psi_o(G) = \beta_G(E^{(G)})$$

The *Graph-State Functor* for arrows, $\Psi_a$, permit us to transform a covered and complete abstraction of graphs into an abstraction of states that preserves the previous definition for $\Psi_o$:

$$\Psi_a : \nabla \rightarrow \blacktriangle$$
$$\forall A \in \nabla, \Psi_a(A) = (\Psi_o(G^{(A)}), \Psi_o(H^{(A)}), \tau, \kappa) \tag{3.14}$$

where functions $\tau$ and $\kappa$ are defined as[10]:

$$\tau^{(\Psi_a(A))} : \Upsilon \rightarrow \Upsilon$$

$$\forall z \in E^{(G^{(A)})}, (def(\tau^{(\Psi_a(A))}(\Gamma_{e1}(z))) \wedge def(\tau^{(\Psi_a(A))}(\Gamma_{e2}(z))) \wedge$$
$$def(\tau^{(\Psi_a(A))}(\Gamma_{e3}(z)))) \Leftrightarrow def(\varepsilon^{(A)}(z)) \wedge$$

$$\forall Z \subseteq E^{(G^{(A)})}, \forall z \in Z, def(\varepsilon^{(A)}(z)) \Leftrightarrow def(\tau^{(\Psi_a(A))}(\Gamma_e^*(Z)))$$

$$\forall z \in E^{(G^{(A)})} : def(\varepsilon^{(A)}(z)),$$

$$\begin{bmatrix} \tau^{(\Psi_a(A))}(\Gamma_{e1}(z)) = \Gamma_{e1}(\varepsilon(z)) \\ \tau^{(\Psi_a(A))}(\Gamma_{e2}(z)) = \Gamma_{e2}(\varepsilon(z)) \\ \tau^{(\Psi_a(A))}(\Gamma_{e3}(z)) = \Gamma_{e3}(\varepsilon(z)) \end{bmatrix} \wedge$$

$$\forall Z \subseteq E^{(G^{(A)})}, \forall z \in Z, def(\varepsilon^{(A)}(z))$$

$$\tau^{(\Psi_a(A))}(\Gamma_e^*(Z)) = \Gamma_e^* \left( \bigcup_{z_i \in Z} \varepsilon^{(A)}(z_i) \right)$$

$$\kappa^{(\Psi_a(A))} : \Upsilon \rightarrow \Upsilon$$

$$\forall a \in V^{(G^{(A)})}, def(\kappa^{(\Psi_a(A))}(\Gamma_v(a))) \Leftrightarrow def(\nu^{(A)}(a))$$

$$\forall a \in V^{(G^{(A)})} : def(\nu^{(A)}(a)), \kappa^{(\Psi_a(A))}(\Gamma_v(a)) = \Gamma_v(\nu(a))$$

Figure 3.3 sketches all the possible relations established between the *CVAGraph** and *AState* categories through $\Psi$. Restrictions given by equation (3.13) are demonstrated for the *Graph-State Functor* $\Psi$ in Appendix B, therefore $\Psi$ is a functor.

### 3.2.4 Hierarchical Planning with *CVAGraph** and *AState*

We impose a particular restriction to the possible complete, covered *H-graphs* we deal with. For our purposes, any abstraction $A_i = (G, H, \nu, \varepsilon)$ of *CVAgraph** must satisfy:

$$\forall z \in E^{(G)}, def(\varepsilon(z)) \Rightarrow \begin{bmatrix} \Gamma_{e1}(z) = \Gamma_{e1}(\varepsilon(z)) \wedge \\ \Gamma_{e2}(z) = \Gamma_{e2}(\varepsilon(z)) \wedge \\ \Gamma_{e3}(z) = \Gamma_{e3}(\varepsilon(z)) \end{bmatrix} \tag{3.15}$$

---

[10] $\Gamma_v$ is considered to be defined for our purposes for all the vertexes of the graphs that are translated into states through $\Psi_o$.

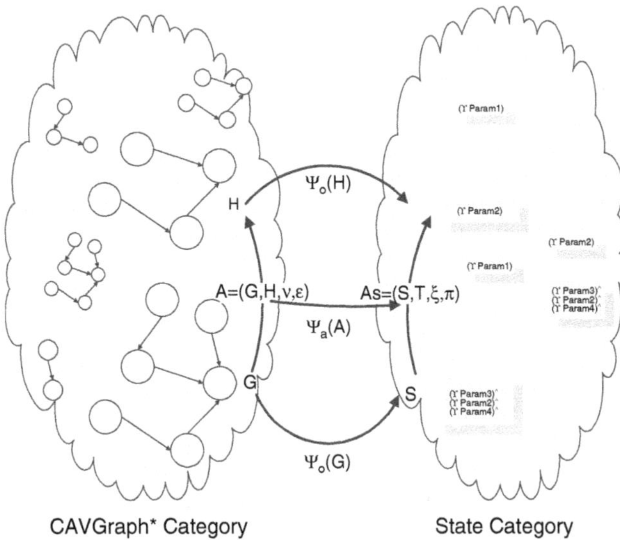


**Fig. 3.3.** The Graph-State Functor. Functor $\Psi$ maps objects and arrows between the *CVAGraph** and *AState* categories. Its inverse is not always defined

$$\forall Z \subseteq E^{(G)}, def(\Gamma_e^*(Z)) \wedge \forall z \in Z, def(\varepsilon(z)) \Rightarrow$$
$$def\left(\Gamma_e^*\left(\bigcup_{z_i \in Z} \varepsilon(z_i)\right)\right) \wedge \Gamma_e^*\left(\bigcup_{z_i \in Z} \varepsilon(z_i)\right) = \Gamma_e^*(Z)$$

That is, the predicate types associated to the edges of a graph remain unaltered when those edges are abstracted.

In order to formalize our Hierarchical Planning approach *HPWA* on the categories previously defined, we need a few more auxiliary definitions.

### Definition 3.2.11 (Parameter-Vertexes Function)
*The* Parameter-*Vertexes Function $\eta$ is a total based on $\Gamma_v$ that transforms a set of parameters from Param into a set of vertexes from $V^{(\Theta^*)}$ as follows:*

$$\eta : power(Param) \rightarrow power(V^{(\Theta^*)})$$
$$\forall P \in power(Param), \eta(P) = \bigcup_{p_i \in P} \Gamma_v^{-1}(p_i)$$

*That is, function $\eta$ produces a set of vertexes from a certain graph that represent a set of given parameters.*

***Remark 3.2.6*** *Note that given a set of vertexes $\eta(P)$, if all are taken from the same graph $G \in \Theta^*$, we can construct the subgraph $G' \subseteq G$, such that:*

$$\left[V^{(G')} = \eta(P)\right] \wedge \left[E^{(G')} = \bigcup_{z_i \in E^{(G)}:ini(z_i)\in\eta(P)\wedge ter(z_i)\in\eta(P)} z_i\right] \tag{3.16}$$

*Thus, from a graph $G$ that represents world information at a certain abstraction level, we can produce a subgraph that only models information related to a given set of parameters $P$.*

**Definition 3.2.12 (Composition Chain of Abstractions)** *A Composition Chain of Abstractions (arrows from either AGraph or AState) is a finite sequence of arrows connected by the corresponding composition function. In AGraph, a composition chain of abstractions is:*

$$\Xi_g = \{A_1 \Diamond A_2 \Diamond \ldots \Diamond A_w\}, \ where$$

$$A_1 = (G_0, G_1, \mu_1, \varepsilon_1),$$

$$A_2 = (G_1, G_2, \mu_2, \varepsilon_2), \ldots,$$

$$A_w = (G_{(w-1)}, G_w, \mu_w, \varepsilon_w)$$

*which involves $w$ arrows ($w$ is said to be the length of the composition). A composition chain of abstractions in AGraph is also called a H-Graph (recall Chap. 2). On the other hand,*

$$\Xi_s = \{A_{s1} \Diamond A_{s2} \Diamond \ldots \Diamond A_{sw}\}, \ where$$

$$A_{s1} = (S_0, S_1, \xi_1, \pi_1),$$

$$A_{s2} = (S_1, S_2, \xi_2, \pi_2), \ldots,$$

$$A_{sw} = (S_{(w-1)}, S_w, \xi_w, \pi_w)$$

*is a composition chain of abstractions of AState involving $w$ arrows.*

**Remark 3.2.7** *Given a composition chain of $w$ abstractions of $CV\,Agraph^*$, that is, given a covered, complete H-graph, there exists a composition chain of $w$ abstractions of AState, $\Xi_s$, that satisfies:*

$$\forall A_i \in \Xi_g, \ [\Psi_a(A_i) = A_{s_i}]$$

*That is, given a composition chain of abstractions of $CV\,Agraph^*$, a corresponding composition of abstractions of AState can be constructed by using the functor $\Psi$.*

As commented before, $\Psi$ establishes a correspondence between $CV$ $Agraph^*$ and $AState$. As described further on, our HPWA approach is aimed to detect and discard world elements (at any abstract level) which are irrelevant with respect to the task at hand. Such an elimination process requires to compute the part (subgraph) of the world information at a given abstraction level of $CV\,AGraph^*$ which is relevant for a plan, and thus, a mechanism to induce that subgraph from a set of relevant world elements is needed.

**Definition 3.2.13 (Extended Abstraction of Planning States)** *The extended abstraction of planning states relaxes the restrictions imposed in definition 3.2.5 in order to construct more general abstractions. This function serves to abstract planning states that can not be directly constructed through the functor $\Psi$, although their parameters stem from a set of vertexes of a graph of $CVAGraph^*$.*

*For clarity sake, we firstly define the* extended abstraction of logical predicates, $EAP$. *Formally, given an abstraction in $CVAGraph^*$, $A = (G, H, \nu, \varepsilon)$ and its corresponding abstraction in AState $A_s = (Q, S, \xi, \pi)$ (constructed by the $\Psi_a$ functor), let $\Sigma_Q$ ($\Sigma_S$) be the set of all logical predicates involving parameters only from the planning state $Q$ ($S$). The extended abstraction function for logical predicates, $EAP_{A_s}$, applied to a predicate $p$ yields another predicate $p'$ such that:*

$$
\begin{gathered}
EAP_{A_s} : \Sigma_Q \to \Sigma_S \\
\forall p \in \Sigma_Q, EAP_{A_s}(p) = p' \in \Sigma_S : \\
\left[
\begin{gathered}
length(Param^{(p)}) = length(Param^{(p')}) \\
SN(p) = SN(p') \wedge \\
\forall j \in 1..length(Param^{(p)}), Param_j^{(p)} = \pi(Param_j^{(p')})
\end{gathered}
\right]
\end{gathered}
\tag{3.17}
$$

*Based on the function $EAP$, the extended abstraction of planning states, $EAbs$, is formalized as follows. Given an abstraction in $CVAGraph^*$, $A = (G, H, \nu, \varepsilon)$, and its corresponding abstraction in AState $A_s = (Q, S, \xi, \pi)$ (constructed by the $\Psi_a$ functor), let $\mho_Q$ ($\mho_S$) be the set of all possible states whose logical predicates are in $\Sigma_Q$ ($\Sigma_S$). The extended abstraction function for planning states, $EAbs$, is defined as:*

$$
\begin{gathered}
EAbs_{A_s} : \mho_Q \to \mho_S \\
\forall s = (p_1^{(s)}, p_2^{(s)}, \ldots, p_n^{(s)}) \in \mho_Q, \\
EAbs_{A_s}(s) = s' = (p_1^{(s')}, p_2^{(s')}, \ldots, p_n^{(s')}) \in \mho_S : \\
\forall p_i^{(s')} \in s', p_i^{(s')} = EAP_{A_s}(p_i^{(s)})
\end{gathered}
\tag{3.18}
$$

**Definition 3.2.14 (Extended Refinement of Planning States)**
*Now we define the inverse of the previous definition. The extended refinement of planning states yields a refined version of a given state $s$ based on refinement of graphs.*

*For clarity, we first define the extended refinement of logical predicates and then the equivalent definition for states.*

*Given an abstraction in $CVAGraph^*$, $A = (G, H, \nu, \varepsilon)$ and its corresponding abstraction in AState $A_s = (Q, S, \xi, \pi)$ (constructed by the $\Psi_a$ functor), let $\Sigma_Q$ ($\Sigma_S$) be the set of all logical predicates only involving parameters from state $Q$ ($S$). The extended refinement function for logical predicates, $ERP_{A_s}$, applied to a predicate $p$ yields a set of predicates $\{p'\}$ such that:*

$$ERP_{A_s} : \Sigma_S \rightarrow power(\Sigma_Q)$$
$$\forall p \in \Sigma_S, \tag{3.19}$$

$$\left[ \begin{array}{c} \forall p' \in ERP_{A_s}(p), \\ \left[ length(Param^{(p')}) = length(Param^{(p)}) \right] \wedge \\ \left[ \forall p' \in ERP_{A_s}(p), SN(p') = SN(p) \right] \wedge \\ \left[ \forall j \in 1..length(Param(p)), Param_j^{(p')} \in \left[\pi^{-1}\right]^{(A_s)} (Param_j^{p}) \right] \wedge \\ \left( \left( \exists x1, x2, \in 1..length(Param(p)) : Param_{x1}^{(p)} = Param_{x2}^{(p)} \right) \Rightarrow \right. \\ \left. Param_{x1}^{(p')} = Param_{x2}^{(p')} \right) \end{array} \right] \wedge \tag{3.20}$$

$$\left[ \bigcup_{p' \in ERP_{A_s}(p)} SP(p') = \bigcup \left[\pi^{-1}\right]^{(A_s)} (SP(p)) \right] \wedge \tag{3.21}$$

$$\left[ \begin{array}{c} \forall j \in 1..length(Param(p)), \forall b \in \left[\pi^{-1}\right]^{(A_s)} (Param_j^{(p)}), \\ \exists p' \in ERP_{A_s}(p) : b \in Param(p') \end{array} \right] \tag{3.22}$$

*Informally, through the definitions given in (3.19) we establish that given a predicate p, its refined predicates must have the same length and predicate name than p. Also we impose that all possible combinations of refined predicates are considered and that several instances of a parameter in p must be refined to the same parameter in each refined predicate.*

*Using the function ERP, the extended refinement of planning states, ERef, is formalized as follows. Given an abstraction in CVAGraph*, $A = (G, H, \nu, \varepsilon)$ and its corresponding abstraction in AState $A_s = (Q, S, \xi, \pi)$ (constructed by the $\Psi_a$ functor), let $\mho_Q$ ($\mho_S$) be the set of all possible states whose logical predicates are in $\Sigma_Q$ ($\Sigma_S$). The extended refinement function for planning states, ERef, is defined as:*

$$ERef_{A_s} : \mho_S \rightarrow power(\mho_Q)$$
$$\forall s, (p_1^{(s)}, p_2^{(s)}, \ldots, p_n^{(s)}) \in \mho_S, \ ERef_{A_s}(s) = \{s'\}$$
$$\text{with } s' = (p_1^{(s')}, p_2^{(s')}, \ldots, p_n^{(s')}) \in \mho_Q : \tag{3.23}$$

$$\left[ \forall p_i^{(s)} \in p, \forall p^* \in ERP_{A_s}(p_i^{(s)}), \exists s' \in ERef_{A_s}(p) : p_i^{(s')} = p^* \right] \wedge \tag{3.24}$$

$$\left[ \begin{array}{c} \forall s' \in ERef_{A_s}(p), \forall p_u^{(s')}, p_w^{(s')} \in s', \\ \left( \begin{array}{c} \exists x1 \in 1..length(Param(p_u^{(s')})) \wedge \exists x2 \in 1..length(Param(p_w^{(s')})) : \\ \pi^{(A_s)}(Param_{x1}^{(p_u^{(s')})}) = \pi^{(A_s)}(Param_{x2}^{(p_w^{(s')})}) \\ Param_{x1}^{(p_u^{(s')})} = Param_{x2}^{(p_w^{(s')})} \end{array} \right) \Rightarrow \end{array} \right] \tag{3.25}$$

## 3.3 Hierarchical Planning Through Plan Guidance

Using the definitions stated in the previous section, the first implementation of our HPWA approach that, for short, will be called *HPWA-1*, is described here. Nevertheless, a reader who has skipped the previous mathematical definitions can still follow the description of our hierarchical planning approaches.

The gist of *HPWA-1* is to solve a task at a high level of world abstraction by running an *embedded planner* (an off-the-shelf classical planner), and then using the resulting abstract plan at the next lower level to ignore the irrelevant elements that did not take part in the abstract plan. This process is repeated recursively until a plan involving primitive actions is constructed that only uses information from the ground level of the hierarchy. HPWA-1 can be also called *Planning through Plan Guidance* since it considers whole plans produced at high levels to discard world elements at lower levels, in contrast to the second method presented in Sect. 3.4, that individually considers each plan action.

A relevant feature of HPWA-1 is that it sets the best bound on the optimality of the plans that can be obtained with any HPWA approach that uses the embedded planner included in the method (similarly as in hierarchical path search by refinement [49]).

Figure 3.5 shows the pseudocode of the HPWA-1 algorithm. Given a complete, covered *H-graph* that satisfy requirements of Sect. 3.2.4, that is, given a composition chain of abstractions of $CV\,Agraph^*$:

$$\Xi_g = \{A_1 \Diamond A_2 \Diamond \ldots \Diamond A_w\}$$

that represents the initial state of the world viewed at different levels of abstraction, and given a *Goal* state expressed with information taken[11] from the ground level of that H-graph[12], noted as $G_0$, the first step of HPWA-1 (and also of HPWA-2) is to compute a composition chain of $w$ *AState* abstractions, $\Xi_s$, through the functor $\Psi$, obtaining a sequence of planning states that represent the initial state at different levels of detail.

---

[11] That is, $\mu(SP(Goal)) \in V^{(G^{(A_1)})}$.

[12] Without generality lose, we assume here that the goal state is given at the ground level of the H-graph (the first graph of the composition chain, that is, $G^{(A_1)}$). Otherwise, the formalization of HPWA is similar, but only considering composition chains of length $w - k$, being $k$ the position of the graph within the H-graph in which the goal is given.

Then, HPWA-1 constructs a sequence of goal states $(Goal, Goal_1, \ldots, Goal_w)$ representing each of them the goal to be achieved in terms of the information taken at the corresponding levels of the *H-graph*. For instance, if $G_w$ represents the most abstracted hierarchical level of the H-graph and $S_w$ the initial planning state obtained via $\Psi_o(G_w)$, $Goal_w$ is the goal to be achieved by the embedded planner. The sequence of goal states is computed by applying the extended abstraction of planning states $EAbs$ to every arrow of the composition chain $\Xi_s$. Thus, $Goal_i$ is obtained by definition 3.18:

$$\forall A_i \in \Xi_s, i > 0, Goal_i = EAbs_{A_i}(Goal_{i-1})$$

Once the sequence of goal states is computed, HPWA-1 starts by running an embedded planner (*Graphplan* or *Metric-FF* in our case) with $Initial = \Psi_o(G_w)$ as the initial state, and $Goal_w$ as the goal to be achieved. With this pair of abstract initial and goal states, and a set of operators $O$ (see Fig. 3.7 for an example), the embedded planner produces an abstract plan at level $w$, that we denote $plan_w$.

Such an abstract plan at level $w$, composed of a sequence of instantiated actions that transforms the initial state into the goal state, will be used to discard irrelevant information at the next lower hierarchical level *(w-1)*. It is important to remark that for our HPWA approach, the graph $G_w$ from which the initial abstract state is produced remains unalterable during the whole planning process (that is, neither edges nor vertexes are added/removed). Only a sequence of intermediate planning states are produced by the embedded planner to achieve the corresponding goal state[13].

Given $plan_w$ that solves the goal $Goal_w$ at level $w$, HPWA-1 discards irrelevant information with respect to the task at hand by obtaining the parameters of $plan_w$, that is[14], the set of symbols (world elements) involved in all instantiated actions of $plan_w$, through the function $\eta(SP(plan_w))$ (see Fig. 3.5). Therefore, $\eta(SP(plan_w))$ yields all vertexes from the abstract level $G_w$ needed to solve $Goal_w$ through $plan_w$.

Through such a set of vertexes, we can induce a subgraph on $G_w$, the so-called $RelevantGraph_w$, that *only* models the (abstract) information that have been used by the task planning process: that is, it does not consider information which is irrelevant for planning the abstract task. In Fig. 3.4 vertexes of the successive *RelevantGraphs* obtained in that way are marked, while irrelevant vertexes are unfilled.

After obtaining each $RelevantGraph_i$, the next step is to plan the task at the lower hierarchical level *(i-1)*, but only using those symbols which are relevant for the task. That is, we only consider information stemmed from

---

[13] Also notice that those intermediate states might not be directly transformable into graphs since $\Psi$ is not a bijection.

[14] A plan, as a sequence of logical predicates, can be also considered as a planning state, and thus, its set of parameters can be computed.

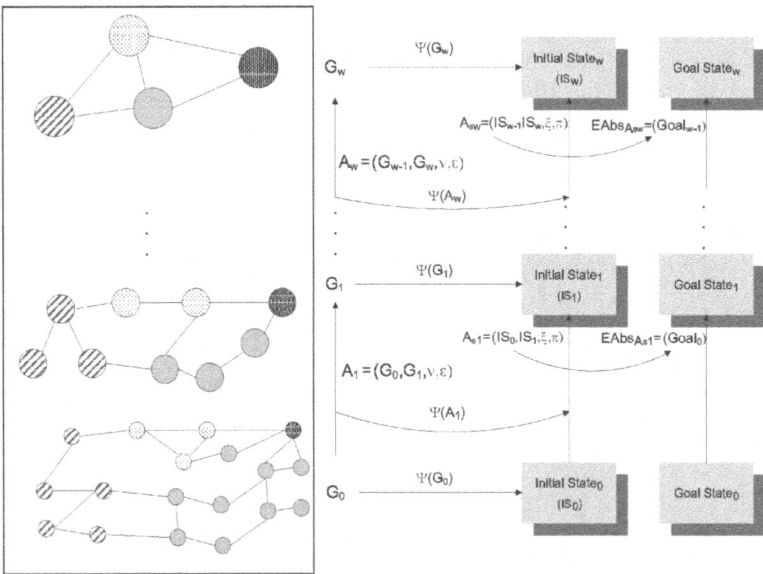


**Fig. 3.4.** Planning scheme upon a hierarchical world model. Given the *Graph-State Functor* ($\Psi$), we can adopt a parallelism from AH-graphs (sequence of graphs and graph abstractions) and planning states, and thus, we can perform abstract planning (using information from abstract levels of the world model), which usually involves less computational effort than planning at the ground level

```
PROCEDURE HPWA-1 (State Goal, AHGraph G):Plan
w=LongestCompositionChain(G)
A=AbstractionCompositionChain(G,w)
Initial= Ψ_o(G_w)
FOR i = w DOWNTO 0
                Goal_i= ExtendedStateAbstraction(Goal,A,i)
                problem-space={Initial,Goal_i,domain-operations}
                plan=EmbeddedPlanner(problem-space)
                RelevantGraph=Induced-Graph(η(SP(plan)))
                IF (i ≠ GroundLevel(G))
                        ReducedGraph=ℓ⁻((H, RelevantGraph, ν, ε))
                        Initial=Ψ_o(H)
                END
END
RETURN (plan)
END
```

**Fig. 3.5.** Pseudocode of the HPWA-1 implementation. A plan constructed from a certain level of an AH-graph serves to reduce the amount of information at lower levels. In the pseudocode, the *ExtendedStateAbstraction* procedure yields the goal state at the *i-th* level of the hierarchy, and $\eta(SP(plan))$ yields the set of vertexes that represent the parameters involved in a plan

the *refinement* of $RelevantGraph_i$ onto $G_{i-1}$. For that, and knowing the abstraction $A_i = (G_{i-1}, G_i, \nu_{i-1}, \varepsilon_{i-1}) \in \Xi_g$, an arrow $A$ in $CVAGraph^*$ is constructed, satisfying that:

$$A = (H, RelevantGraph_i, \nu', \varepsilon') :$$

$$[H \subseteq G_{i-1} \wedge RelevantGraph_i \subseteq G_i]^{15} \wedge$$

$$\begin{bmatrix} \forall z \in E^{(H)} : def(\varepsilon'(z)) \Leftrightarrow def(\varepsilon_{i-1}(z)) \wedge def(\varepsilon'(z)) \Rightarrow \\ \left[(\varepsilon'(z) = \varepsilon_{i-1}(z)) \wedge \varepsilon_{i-1}(z) \in E^{(RelevantGraph_i)}\right] \\ \\ \forall a \in V^{(H)} : def(\nu'(a)) \Leftrightarrow def(\nu_{i-1}(a)) \wedge def(\nu'(a)) \Rightarrow \\ \left[(\nu'(a) = \nu_{i-1}(a)) \wedge \nu_{i-1}(a) \in V^{(RelevantGraph_i)}\right] \end{bmatrix}$$

The dual of this abstraction $A$, that is, the refinement $R_A$, is:

$$R_A = (RelevantGraph_i, H, \mu'_i, \alpha'_i)$$

$R_A$ is always defined, since we are considering complete and covered abstractions, and it yields a graph $H$, such that $H \subseteq G_{i-1}$. Informally, $H$ is the refinement of $RelevantGraph_i$ and only contains the minimal set of symbols required to obtain a plan at level $i$. Therefore, symbols not involved in the generation of a plan at a certain level are no longer considered in the planning process at lower levels.

Once the $RelevantGraph_i$ is refined (into the graph $H$), the planning process continues (until the lowest level of the model is reached[16]), taking $Goal_{(i-1)}$ as the goal state to be achieved now and $\Psi_o(H)$ as the initial state, which is a reduced portion of the original state $IS_{(i-1)}$ (refer to Fig. 3.4 again). If at some step of refinement, no plan can be obtained, backtracking occurs and a new abstract plan must be constructed.

In order to illustrate the described HPWA-1 method, let us consider a relatively simple environment where a mobile robot with a manipulator on board has to perform different tasks involving a given object of the world named "Box". Figure 3.6 shows this hierarchical model of the environment with three zones (a laboratory, a room, and a corridor), the mobile robot itself, and the box to be handled. In this model, a complete, covered *H-graph* with three hierarchical levels has been considered, and thus a composition chain of two abstractions of graphs is defined:

$$A_1 = (G_0, G_1, \nu_1, \varepsilon_1)$$
$$A_2 = (G_1, G_2, \nu_2, \varepsilon_2)$$
$$\Xi_g = \{A_1 \Diamond A_2\}$$

---

[15] We indicate with $H \subseteq G_{i-1}$ that $H$ is a *subgraph* $G_{i-1}$. A subgraph of a graph $G$ is a graph whose vertex and edge sets are subsets of those of $G$.

[16] In the case that the original goal state uses information from a certain level of the *H-Graph*, say $G_k$, different from the ground one, HPWA-1 will stop at that level.

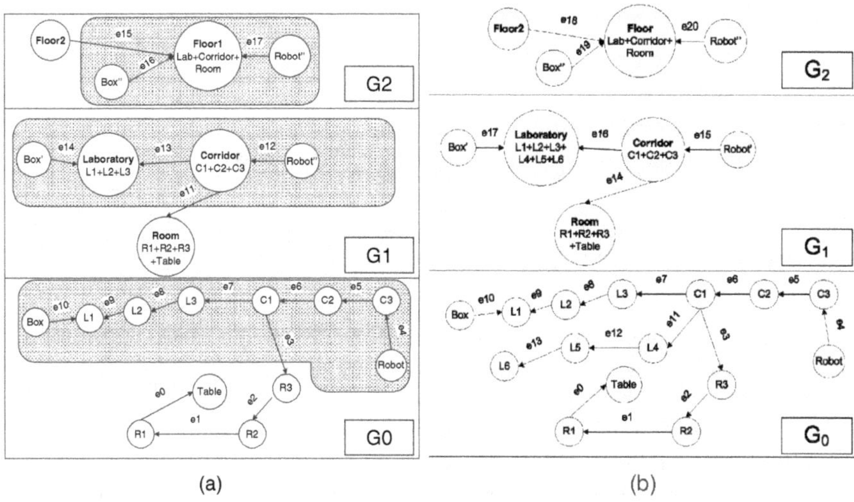


(a)                                        (b)

**Fig. 3.6.** Example of two different hierarchical models of a robot environment. For clarity, from the hierarchical level $G_0$ upwards, each vertex label includes its subvertexes' labels. For example, in (a), the *Laboratory* node is the supervertex of $L1$, $L2$, and $L3$. Also in (a), it is gray-shaded the subgraph of each level that represents the relevant information for the example explained in the text

The ground level of the *H-graph*, $G_0$, has the maximum amount of detail (vertexes represent distinctive locations for navigation and manipulation), the first level, $G_1$, is used for grouping vertexes of $G_0$ into three different areas (which allows the robot to reset odometric error through some sensor matching method [98, 99]), and finally, level $G_2$, represents the world with the minimum amount of detail.

In this example, given a graph $G \in \{G_0, G_1, G_2\}$, we define the following three sets of edges:

$$Z = \{e_1, e_2, e_3, e_5, e_6, e_7, e_8, e_9, e_{11}, e_{13}, e_{15}\},$$
$$X = \{e_0, e_{10}, e_{14}, e_{16}\},$$
$$Y = \{e_4, e_{12}, e_{17}\}$$

such that:

- $\Gamma_e(Z) = (location, location, nav)$ : $location, nav \in \Upsilon$
- $\Gamma_e(X) = (object, location, at)$ : $object, location, at \in \Upsilon$
- $\Gamma_e(Y) = (robot, location, at\text{-}robot)$ : $robot, location, at\text{-}robot \in \Upsilon$

That is, $\Gamma_e$ defines, in this example, three different types of arcs: (1) those arcs that indicate the robot possibility of navigating (*nav*), and thus, the type of their initial and terminal vertexes as *locations*; (2) arcs that represent the position of objects (*at*), being the object represented by the initial vertex, and the location where it is, by the terminal vertex. And finally, (3) the arc (in our

```
(operator
 GO (params (< x > LOCATION) (< y > LOCATION) < z > ROBOT)
               (preconds (at-robot < z > < x >) (nav < x > < y >) )
               (effects (del at-robot < z > < x >) (at-robot < z > < y >)))

(operator
 PICKUP (params (< x > OBJECT) < y > LOCATION < z > ROBOT))
               (preconds (at-robot << z > < y >)(at-object < x > < y >))
               (effects (held < x >)(del at-object < x > < y >)))
```

**Fig. 3.7.** Definition of the operators used in the example of HPWA-1. Operator effects can add or remove logical predicates from the planning state

example there is only one since there is only one robot) that represents the position of the robot (*at-robot*) for which its initial vertex is typed as *robot*, while the terminal one is as *location*.

$\Gamma_v$, which is the function that transforms vertexes into parameters, simply takes the vertex label (shown in Fig. 3.6 inside the vertexes) as a constant string. This is a simple translator function that imposes that vertex labels must be unique across all levels of the *H-graph*.

In this scenario, planning a task like for example "pickup Box" implies to search a plan to achieve a ground goal state $\{(held\ Box)\}$ (as the postcondition of the "pickup" operator defined in Fig. 3.7). According to HPWA-1, this task is planned as follows.

First, the composition chain of *AState* as well as the sequence of initial states for planning are computed through the functor $\Psi$. Thus:

$$IS_0 = \Psi_o(G_0), IS_1 = \Psi_o(G_1), IS_2 = \Psi_o(G_2)$$
$$A_{s1} = \Psi_a(A_1), A_{s2} = \Psi_a(A_2)$$

Through the abstractions of the composition chain $\Xi_s = \{A_{s1}\Diamond A_{s2}\}$, the goal state, $\{(held\ Box)\}$, can be expressed at every level of the hierarchy by applying the *EAbs* function. For this example it is:

$$Goal_0 = \{(held\ Box)\}, Goal_1 = \{(held\ Box')\}, Goal_2 = \{(held\ Box'')\}$$

With these goals to be achieved at each level, HPWA-1 starts by planning at level $G_2$ the abstract goal $Goal_2 = \{(held\ Box'')\}$:

- *Planning at the universal level $G_2$*
  The initial state $IS_2$ for level $G_2$ is $\Psi_o(G_2)$, which, in this example, according to the type of edges, contains the following predicates:

$$\Psi_o(G_2) = \{\texttt{(at Box'' Floor1), (at-robot Robot'' Floor1),}$$
$$\texttt{(object Box''), (location Floor1), (robot Robot''),}$$
$$\texttt{(location Floor2), (nav Floor1 Floor2)}\}$$

Since $e_{15} \in Z$, $e_{16} \in X$, and $e_{17} \in Y$, that is, $e_{15}$ is a navigational arc, $e_{16}$ indicates the location of an object, and $e_{17}$ indicates the location of the robot (see Fig. 3.6).

The embedded planner is run with this initial state to solve the abstract goal $Goal_2 = \{(held\ Box'')\}$, finding the following solution:

$$Plan_2 = (\text{PICKUPBox''Floor1Robot''})$$

Notice that in $CVAGraph^*$, due to the manner in which graphs are translated into states, some abstractions may lead to a losing of relevant information for the planning process. For instance, in this example note that if the vertex $Robot$ from level $G_0$ would have been abstracted to the vertex $Corridor$ (see Fig. 3.6), information about the location of the vehicle would be lost at upper levels, and thus, the planning process would have been impossible. We assume that the world information (provided by edges) relevant for the goal to be solved is maintained along the abstractions.

- *Planning at level $G_1$*

  Now, the symbolic information involved in $Plan_2$ is computed through function $\eta$. In this case,

$$\eta(SP(Plan_2)) = \{\texttt{Box''}, \ \texttt{Floor1}, \ \texttt{Robot''}\}$$

Then, we compute the subgraph of $G_2$ induced by $\eta(SP(Plan_2))$, resulting a graph called $RelevantGraph_2$ (shown as a gray-shaded area at level $G_2$ in Fig. 3.6a). For this example, $RelevantGraph_2$ does not include the symbol $\texttt{Floor2}$, since it is irrelevant to solve the task (it does not belong to $\eta(SP(Plan_2))$).

$RelevantGraph_2$ is then refined down to level $G_1$, where only subvertexes of its vertexes are considered, and thus, only world elements which are relevant when solving the task are refined. In this particular case no information is ignored at level $G_1$, since the unique discarded vertex $Floor2$ has no subvertexes. Thus, at level $G_1$, the initial state is:

$\Psi_o(G_1)$={(object Box'), (robot Robot'), (location Laboratory)
            (location Corridor), (location Room),
          (at Box' Laboratory), (at Robot' Corridor),
        (nav Corridor Laboratory), (nav Corridor Room)}

provided that $\{e_{11}, e_{13}\} \in Z$, $e_{14} \in X$, and $e_{12} \in Y$.
The goal state at this level is $Goal_1 = \{(held\ Box')\}$. A solution for this goal provided by the embedded planner is:

$$Plan_1 = (\text{GO Corridor Laboratory}), (\text{PICKUP Box' Laboratory})$$

Observe that the world element "Room" does not appear in the plan (it is irrelevant for the task ) and therefore, the $RelevantGraph_1$ (shown shaded at level $G_1$ in Fig. 3.6a) becomes a subgraph of $G_1$. At the next lower level (ground level), all its subvertexes will be discarded.

- *Planning at the ground level $G_0$*

  Finally, at the ground level, the predicates that model the initial state come from the refinement of $RelevantGraph_1$, that does not contain the

*Room* vertex. Thus, the initial state is obtained through the refinement graph $H$ of *RelevantGraph$_1$*:

$$\Psi_o(H)=\{\texttt{(object Box)}, \texttt{(object Table)}, \texttt{(robot Robot)},$$
$$\texttt{(location L1)}, \texttt{(location L2)}, \texttt{(location L3)}, \texttt{(location C1)},$$
$$\texttt{(location C2)}, \texttt{(location C3)}, \texttt{(at Box L1)}, \texttt{(at Robot C3)},$$
$$\texttt{(nav C3 C2)}, \texttt{(nav C2 C1)}, \texttt{(nav C1 L3)}, \texttt{(nav L3 L2)}, \texttt{(nav L2}$$
$$\texttt{L1)}\}$$

Given that $\{e_5, e_6, e_7, e_8, e_9\} \in Z$, $\{e_0, e_{10}\} \in X$, and $e_4 \in Y$. Observe that no vertex within the *Room* is included in this state, hence reducing the computational cost of planning[17].

The goal state at the ground level is $\{$*(held Box)*$\}$, which is finally solved as:

$$Plan_0= \text{(GO C3 C2),(GO C2 C1),(GO C1 L3),(GO L3 L2),(GO L2 L1),}$$
$$\text{(PICKUP Box L1)}$$

These actions are primitive actions that the mobile robot can execute, so this is the final plan for the original task at hand.


## 3.4 Hierarchical Planning Through Action Guidance

*Planning through Action Guidance* (HPWA-2 for short) differs from the previously commented HPWA-1 in that HPWA-2 uses the abstract plans not only to ignore irrelevant elements, but also to guide refinement into lower levels. This can lead, in normal situations, to a higher reduction in computational cost than HPWA-1; however, due to the poorer look-ahead capability of this refining process, HPWA-2 is more sensitive to backtracking than HPWA-1 .

Broadly speaking, HPWA-2 tries to solve individually each action of an abstract plan (that is, an instantiated planning operator). To do that, postconditions of every action of a plan are considered as goal states to be sequentially planned at the lower level of the hierarchy. Thus, a particular refinement of the parameters involved in each action postcondition must be selected. That is, a subvertex of those vertexes from the *H-graph* that represent the parameters of a postcondition $p_i$ (given by $\eta(SP(p_i))$) must be selected.

Figure 3.8 sketches the HPWA-2 algorithm. HPWA-2 starts from an abstracted plan, taking each action individually in order to refine it (to obtain a subplan for every of them) at the next lower level. The refinement of an action $a_i$, that is, the process of refining its parameters, is carried out

---

[17] This is only an illustrative example: more extensive results on the reduction in computational cost are presented in Sect. 3.6.

```
PROCEDURE HPWA-2 (State Goal, AHGraph G):Plan
w=LongestCompositionChain(G)
A=AbstractionCompositionChain(G,w)
Initial= Ψ(G_w)
FOR i = w DOWNTO 0
        Goal_i= ExtendedStateAbstraction(Goal,A,i)
        problem-space={Initial,Goal_i,domain-operations}
        plan=EmbeddedPlanner(problem-space)
        action=FirstActionPlan(plan)
        WHILE (action!=NULL)
                r-action=ActionRefinement(action)
                IF (r-action==NULL) action=PreviousActionPlan(plan)
                ELSE
                        Goalaux=postcondition(r-action)
                        actionGraph=Induced-Graph(η(SP(Goalaux)))
                        ReducedGraph=ℓ⁻((H, actionGraph, ν, ε))
                        Initial=Ψ(H)
                        problem-space-aux={Initaux,Goalaux,domain-operations}
                        planaux=EmbeddedPlanner(problem-space-aux)
                        IF (planaux!=NULL)
                                newplan=Concatenate(newplan, planaux)
                                Do(action)
                                action=NextActionPlan(plan)
                        ELSE
                                newplan=EliminateLast(newplan)
                                action=PreviousActionPlan(plan)
                                Undo(action)
                        END
                END
        END
plan=newplan
END RETURN (plan)
```

**Fig. 3.8.** Pseudocode of HPWA-2. HPWA-2 successively plans abstract actions in lower abstract graphs. The refinement of actions will determine the success of planning, since a wrong selection produces backtracking (*Undo(action)*)

randomly[18] in our current HPWA-2 implementation, and thus, the postcondition of a refined action $a_i$, is computed by choosing a random subvertex of each vertex in $\eta(\text{postconditions}(SP(a_i)))$.

The advantage of refining the abstract plan in this way is that the amount of information necessary to plan a unique abstract action is generally much smaller than the information required to plan with all relevant subelements. However, backtracking in HPWA-2 is more likely to occur. We illustrate the operation of this method with the world of Fig. 3.6b, where the ground level is a slight variation of example 3.6a. Since the work of HPWA-2 is similar to HPWA-1 in some aspects, in the following only the relevant differences of HPWA-2 are described.

---

[18] Heuristics could be used, like selecting a border subvertex, that is a vertex $a$ directly connected to a vertex $b$ such that their supervertexes are different, or the subvertex with the highest order, that is, the one which is connected the most to other ones, etc. Other techniques based on constraint propagation are also applicable to guide the subvertex selection process [10].

We start from the abstract plan $Plan_1$ (the first steps of HPWA-2 are identical to HPWA-1) that we repeat here for convenience:

$Plan_1$=(GO Corridor Laboratory), (PICKUP Box' Laboratory Robot')

HPWA-2 refines individually every action that makes up the abstract plan. In this example, the first action, *(GO Corridor Laboratory)*, has the postcondition *(at-robot Laboratory)*, and thus, a refinement of its parameters is chosen for becoming the goal to be solved at the next lower level $(G_0)$. Such a refinement is calculated by choosing a subvertex of those vertexes from the *H-graph* that correspond to the parameters of the postcondition. That is,

$$A = (G_0, G_1, \nu, \varepsilon)$$
$$\eta(SP(\{(\text{at-robot Laboratoy})\}))=\text{Laboratory}$$
$$\nu^{-1}(Laboratory) = B \in V^{(G_0)}$$

The set $B$ contains all subvertexes of vertex *Laboratory*. Thus, we have to choose a parameter $x$ such as:

$$x \in \bigcup_{a \in B} \Gamma_v(a)$$

That is, to obtain a refined version of the postcondition of the action, we have to choose a subvertex of the vertex that represents the symbol *Laboratory* in the *H-graph*. The procedure to choose such a subvertex $x$ can be implemented heuristically, randomly, or in another way, but all of these possibilities may produce wrong elections that make the plan to fail and then, the algorithm must backtrack to select another subvertex. For example, if $x = L6$ (the refined post-condition will be *(at-robot L6)*) it will be impossible to plan the next action, whose unique refinement is (held Box), since there not exists a navigational arc to go back to $L1$ where the box is. In this case, HPWA-2 must choose another subvertex until it finds a feasible plan.

If another vertex is selected for the first action, i.e. $x = L_2$ (the goal state becomes *(at-robot L2)*), a solution for the action *(GO Corridor Laboratory)* would be:

$Plan_{action1}$=(GO C3 C2), (GO C2 C1), (GO C1 L3), (GO L3 L2)

After applying the instantiated postcondition of this partial solution *(Do(action)* in the pseudocode of Fig. 3.8), the initial state for planning the next action is:

```
Initial={(at Box L1), (at-robot Robot L2),
    (object Box), (robot Robot), (location L1),
 (location L2), (location L3), (nav L3 L2), (nav L2 L1)}
```

while the goal to be planned is *held Box*. In this planning state, the next action can be solved successfully as:

$$Plan_{action2}= (GO\ L2\ L1),\ (PICKUP\ Box\ L1\ Robot)$$

The final result of HPWA-2 is the concatenation of both subplans $(Plan_{action1} + Plan_{action2})$, hence solving the original given goal.

## 3.5 Anomalies in Hierarchical Planning

In any classical task planner there exist some anomalies that must be studied. In this section, two of them are briefly mentioned in the context of the HPWA approach. The *Sussman's Anomaly* consists of losing part of the achievements (goals) previously stated by the planner when it intends to achieve new ones [142]. This problem is not directly related to abstraction, although it could appear in any abstraction-based planner. The essential condition under which a given planner is sensitive to this anomaly is that the goals of a plan are achieved separately and considered independently [142]. Since the HPWA methods use an embedded planner that solves a given goal in only one, atomic operation, no separation of goals is carried out.

The reader may think that HPWA-2 suffers from this anomaly since it individually plans refined actions (subgoals) of a plan. But given that HPWA-2 starts from a previously computed (abstract) plan generated by an embedded planner insensitive to this anomaly, HPWA-2 will neither be affected from the Sussman's anomaly. The unique consequence of refining actions independently is the need of backtracking when a wrong refinement makes that planning the next action is impossible. Thus, summarizing, as long as the embedded planner does not suffer from the Sussman's Anomaly, HPWA does not either.

A slightly different anomaly that can be present in any abstraction planner appears when an abstract plan loses its truth at refinement, that is, the refinement of an abstract plan makes false some achieved abstract goal [83]. Assuming the restrictions followed in this chapter, that is, considering complete, covered *H-graphs* and the defined functor $\Psi$, the HPWA methods are not affected from this anomaly. When such restrictions are not considered, some hierarchies may lead to the impossibility of planning certain goals. For instance, following the planning example of Sect. 3.3 we can transform the complete, covered *H-graph* depicted in Fig. 3.6 into a non-complete hierarchy by eliminating the edge $e7$ at $G_0$. In this case, the action *(Go Corridor Laboratory)* of $Plan_1$ can not be solvable at the ground level, while it is at level $G_1$.

In the next section we experimentally compare both hierarchical planning approaches with respect to other hierarchical and non-hierarchical planners, like ABSTRIPS, Graphplan, and Metric-FF.

## 3.6 Experimental Results

This section describes experimental results of the HPWA hierarchical methods
described in this chapter, when applied to larger and more realistic worlds than
that of Fig. 3.6.

Planning complex tasks in an office environment, as the one depicted in
Fig. 3.9, is a challenge for any intelligent robot. This type of scenario may be
composed of a large number of rooms and corridors where a mobile robot
(perhaps equipped with a robotic arm) can perform a certain number of
operations.

We have tested our hierarchical planning approaches within this environment where a mobile robot is intended to find a plan to deliver envelopes
to different offices. The requested envelope can be stacked, so the robot may
need to unstack other envelopes on top of it before taking it. Once the mail
is delivered, the robot must go to a specific place to recharge energy (the
supply room). For the sake of simplicity we assume that all doors are opened.
The definitions of the operators used in the tests presented in this section
(shown in Appendix C) are more complex than those shown in Fig. 3.7, since
they include unstack objects and recharge energy operators. The robot world
hierarchy used by our hierarchical planning methods is shown in Fig. 3.10,
and resembles the human cognitive map [90]. The ground level contains the
maximum amount of detail: every distinctive place, environment objects, and
their relations. The first level groups places and objects to form the human
concept of rooms and corridors. Upper hierarchical levels represent groups of
rooms or areas, and finally, the universal graph contains a single vertex that
represents the whole environment. As commented further on (in Chap. 5),
this world hierarchy might be automatically constructed with respect to some
criteria.

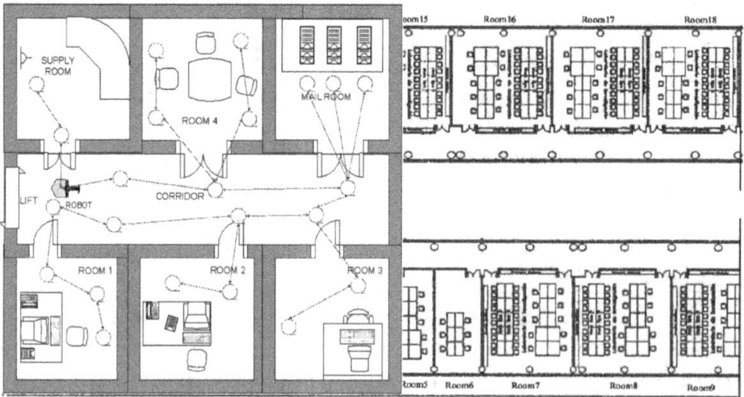


**Fig. 3.9.** A large-scale office environment composed of many rooms. On the left, a
detailed portion of the whole environment is zoomed. Part of the hierarchical ground
level has been overprinted to clarify this partial view

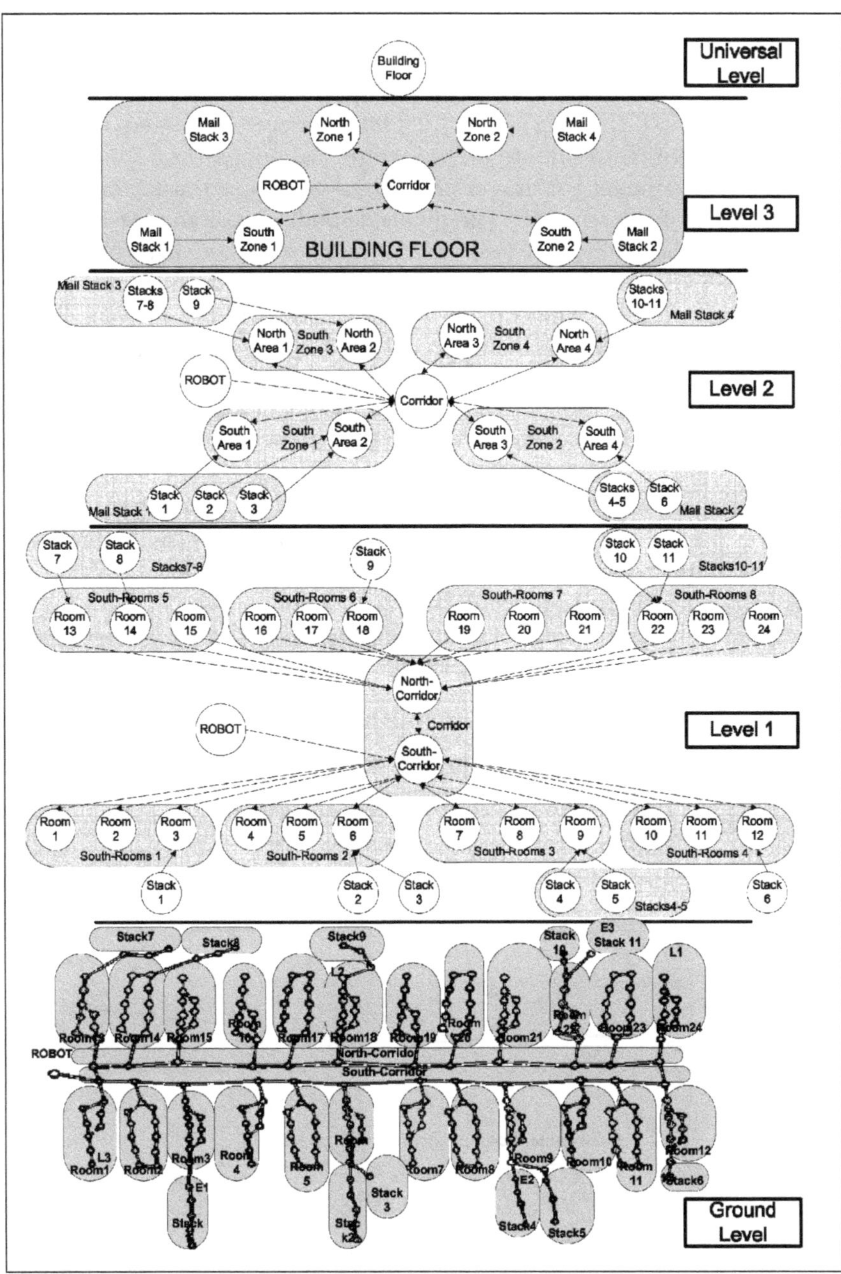


**Fig. 3.10.** Hierarchical representation of the robot environment. Vertexes into a given shaded region are abstracted to the same supervertex at the next higher level. At the ground level symbols (labelled with "E" for envelopes and "L" for destinations) are grouped into vertexes that represent rooms, these ones into areas and finally, a vertex represents the whole environment at the universal level

Within this scenario, we compare the Graphplan and Metric-FF conventional planners to our implementations of HPWA that uses them as embedded planners . Also, an implementation of Graphplan that uses the hierarchies produced by the ABSTRIPS algorithm [128] has also been evaluated, in order to compare our planning results to other hierarchical approaches.

The first experiment consists of planning a single fixed task while varying the complexity of the scenario. The planned robot task is to pick up a given envelope from the mail room (see Fig. 3.9). This task requires the robot to navigate to this room, unstack other envelopes (if needed), and then pick up the requested envelope. The complexity of the scenario ranges from 6 to 48 rooms. Figure 3.11 shows a comparison of the computational costs of planning using HPWA-1 and HPWA-2 with Graphplan as embedded planner against the Metric-FF and Graphplan conventional planners alone, and against the hierarchical version of Graphplan with ABSTRIPS. Observe that, without using abstraction, the computational cost grows exponentially with the number of rooms; however both HPWA-1 and HPWA-2 spend a constant time since the new rooms added to the environment, which are irrelevant for the task, are promptly discarded by the abstract planning process.

A second experiment is aimed to test the method in a medium-complexity environment (with 24 rooms, 7 of them are "mail rooms", that is, they contain stacks of envelopes) where six arbitrary tasks are planned. The first three tasks

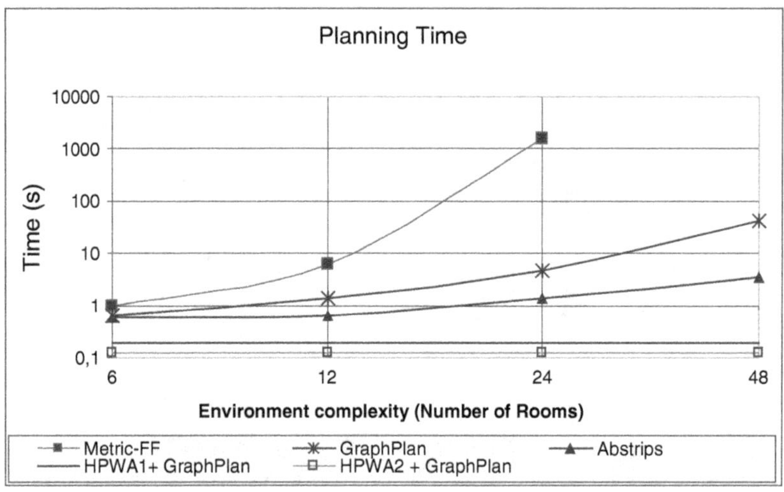


**Fig. 3.11.** Planning a single fixed task in an increasing complexity environment (ranging from 6 to 48 rooms). While the CPU planning time grows exponentially using the conventional planners considered here (hierarchical and non-hierarchical), both HPWA methods exhibit a constant CPU time, since irrelevant information for the task is discarded. This plot only shows the time spent by planners: neither pre- or post-processing, nor communication burdens have been added. No result is shown for Metric-FF for 48 rooms due to the large computational resources demanded

| Tasks #1, #2, #3 | Take Envelopes $E1$, $E2$, or $E3$ |
| Tasks #4, #5, #6 | Carry Envelopes $E1$, $E2$, or $E3$ to locations $L1$, $L2$, or $L3$ |

**Fig. 3.12.** The six tasks planned for the second experiment. Three different objects (envelopes) and three locations of the environment have been chosen at random to test the hierarchical planning methods versus conventional planning. Please, see Fig. 3.10 to find these objects and locations in the ground level of the AH-graph

are "take envelope $E$" and the others "carry envelope $E$ to location $L$" (see Fig. 3.10). The elements involved in the tasks ($E$ and $L$) have been selected arbitrarily from the robot world. All tasks involve navigation and manipulation operations, and some of the goals are reached in more than 40 actions. The whole robot world is composed of more than 250 distinctive places, 30 different objects that the robot can manipulate, and all the navigation and manipulation relations existing between these elements.

Figure 3.13 shows the results of this experiment using HPWA-1 and HPWA-2 with Graphplan as embedded planner, HPWA-1 with Metric-FF, a hierarchical version of Graphplan using the hierarchies produced by ABSTRIPS, and the conventional planners alone (Graphplan and Metric-FF). Each chart shows the results of planning one task out from the ones shown in Fig. 3.12. It is clear the computational benefit of hierarchical planning through world abstraction against both non-hierarchical planning and ABSTRIPS. Also, notice that planning time is not shown for the last three tasks (Task 4, Task 5 and Task 6) for Graphplan planner and ABSTRIPS, because Graphplan was not able to find a plan due to the large computational resources demanded and ABSTRIPS fails in finding a correct plan that solves the tasks due to violation of previously achieved preconditions when refining a plan (the same problem is reported in [82]). In all these plots we only consider the time spent by the embedded planner, without taken into account the pre-processing time taken by the hierarchical manipulation. Notice how HPWA-2 performs better than HPWA-1 in all tasks, except for the first two, because of the backtracking burden.

Figure 3.14 compares the total time (which accounts also for pre-processing and communications) of HPWA-1 versus HPWA-2 when using Graphplan as embedded planner. In this case, HPWA-2 performs more inefficiently since backtracking increases the number of plans to solve. Also, observe how the total time of HPWA-1 is quite similar to its planning time (Fig. 3.13), since the additional cost of this method is negligible.

## 3.7 Conclusions

In AI literature, task planning has not dealt with large domains, which is a common situation in mobile robotics. In these cases, task planning, even with the most modern improvements, can exhibit a high computational cost, even

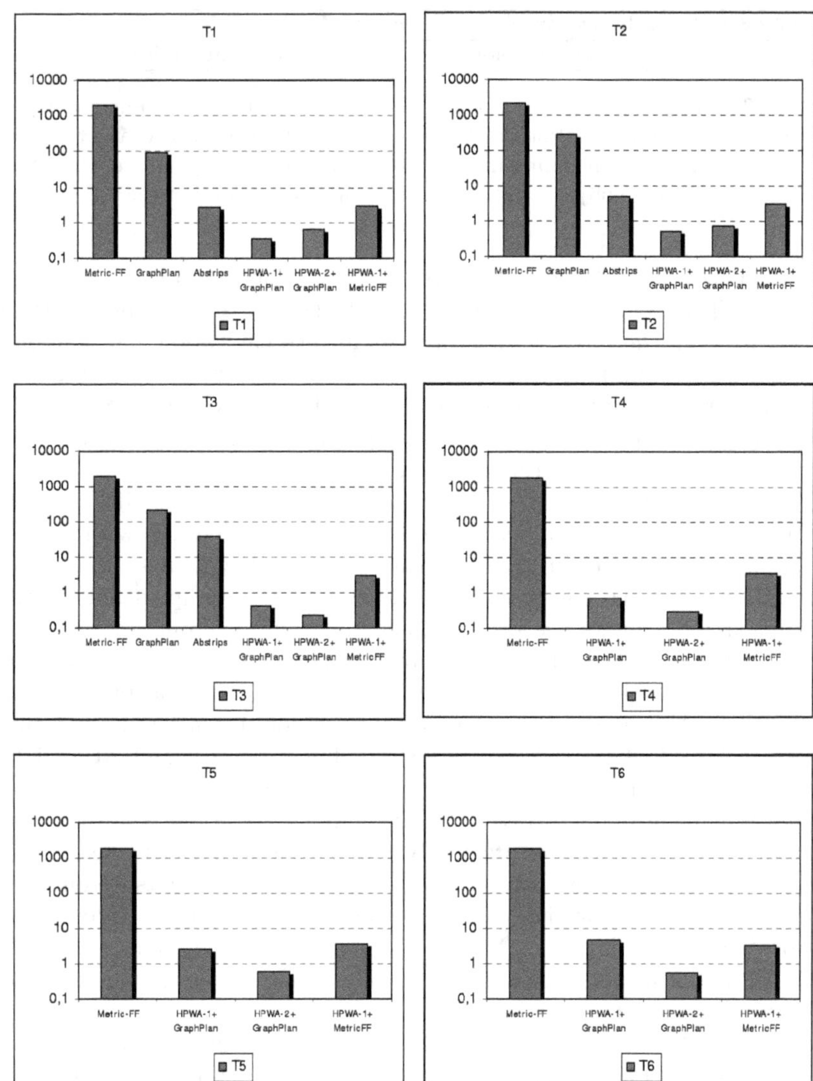

**Fig. 3.13.** Planning time for the proposed tasks (in seconds) on a Pentium IV at 1.7 MHz with 512 Mbytes of RAM. Both non-hierarchical and hierarchical planners exhibit worse CPU time than HPWA methods. Moreover, HPWA-2 exhibits the best planning time for all tasks except for the first two ones due to backtracking situations. Notice that the planning time is not shown for Graphplan and ABSTRIPS in the last three tasks, because they were not able to end up with a solution. In all these plots we consider only the time spent by the embedded planner, without taken into account other pre-processing times

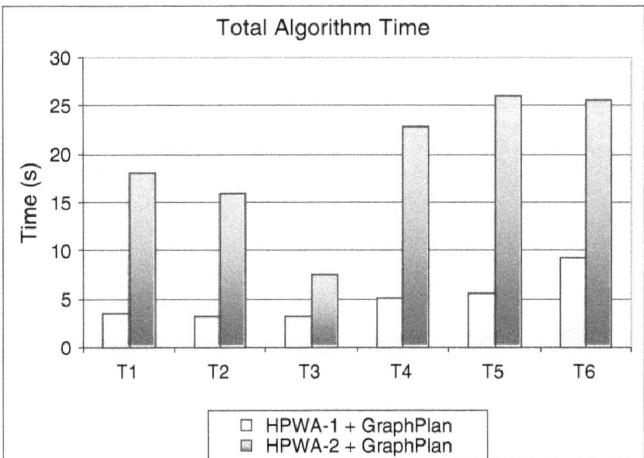



**Fig. 3.14.** Total time spent by our HPWA-1 and HPWA-2 implementations with Graphplan as embedded planner. Time of HPWA-2 is higher than HPWA-1 due to the extra pre-processing involved in backtracking. In spite of this apparently bad result, HPWA-2 performs better than other approaches in Fig. 3.13, even when they do not include the pre-processing and communication burden considered in this plot

can become intractable (as shown in Fig. 3.13). This chapter has introduced a new scheme for task planning that takes advantage of a hierarchical arrangement of the robot model of the environment. This approach performs planning at a high level of abstraction of the world and then, recursively, refines the resulting abstract plan at lower levels, ignoring irrelevant world elements that do not take part in the abstract plan.

HPWA has been stated and formalized through Category Theory upon the categories of abstractions of graphs described in Chap. 2. In particular, we have described two implementations of HPWA that embed other existing planners (Graphplan and Metric-FF) to solve planning at each level of abstraction of the world. Thus, we can benefit from any other kind of improvements on the embedded task planner. We have shown how our two implementations of HPWA have performed better than the embedded planners alone and other hierarchical approaches such as ABSTRIPS.

This good performance is tightly coupled with the use of adequate hierarchies of abstraction. Adequate hierarchies can be obtained by using algorithms for the automatic construction of abstractions based on the task-driven paradigm [49] or the one described in Chap. 5, which is based on an evolutionary approach.

# Mobile Robot Operation with Multi-Hierarchies

> *Science is organized knowledge. Wisdom is organized life.*
> Immanuel Kant (1724–1804)
>
> *Minds are like parachutes. They only function when they are open.*
> Sir James Dewar, scientist (1877–1925)

In this chapter we study the benefits of using a multi-hierarchical symbolic model of the environment in robotic applications. As presented in Chap. 3, a hierarchy, that is, a sequence of graphs and abstractions of graphs, can improve efficiency in certain robot operations like task planning. In [50] it has been stated how these benefits are also present in other tasks, like routing. Here we follow the same inspiration and endow a mobile robot with different hierarchies upon a common ground level (obtaining a *multi-hierarchical model*) in order to improve not only task-planning, but also other operations, such as localization and user communication, simultaneously. Such a model may imply the necessity of translating symbols from one hierarchy to another, especially to the one used for human communication. This chapter also proposes a mechanism that copes with that translation.

## 4.1 Introduction

The human use of hierarchical symbolic structures for dealing with our daily life is well accepted ([68, 71, 89, 90]). As commented in previous chapters, humans use *abstraction* to reduce the huge amount of information gathered by our senses, hence constructing a hierarchy of concepts. Psychological evidences of that can be found in [72, 106]. Also, we have empirically demonstrated in Chap. 3 the benefits of using a hierarchical arrangement of spatial information to reduce the computational effort of task planning.

C. Galindo et al.: *A Multi-Hierarchical Symbolic Model of the Environment for Improving Mobile Robot Operation*, Studies in Computational Intelligence (SCI) **68**, 65–83 (2007)
www.springerlink.com                    © Springer-Verlag Berlin Heidelberg 2007

Moreover, it seems that humans also use multiple abstraction [116] (multiple hierarchies of abstraction built upon the same ground information) for improving our adaptation to different environments and different tasks. That is, having multiple hierarchies allows us to select the most convenient one to perform each operation. Some interesting ideas arise when exploring this multi-hierarchical paradigm [49]:

- Hierarchies of concepts allow humans to perform operations more efficiently than using non-hierarchical, flat data.
- The hierarchy of concepts that is the best one for a given operation depends on the sensorimotor apparatus of the agent and on the particular environment where it operates.
- The hierarchy of concepts that is good for executing efficiently a given operation may be not so good to execute a different one.
- Thus, it is desirable to construct more than one hierarchy upon the same set of ground elements if, for example, more than one task is to be performed, or when the agent has to operate in very complex or different environments.

How do we use multiple abstraction in our lives? Let's consider our different points of view when performing different tasks. For instance, when we drive our car we consider junctions and streets with their directions, traffic signals (like speed limits and wrong way signals), etc. For this car driving activity, we plan paths considering such information; we group streets following urban districts and we usually take into account our previous experience to avoid traffic jams. In contrast, when we are pedestrians, our hierarchical space model changes substantially. Although it contains the same ground space elements, we neither need to take care of traffic signals (only traffic lights for pedestrians), nor to worry about traffic jams. We, as pedestrian, may group (abstract) spatial concepts, like streets, with respect to different criteria, in order to plan routes, for example, by choosing pedestrian streets as often as possible.

In general, the same physical objects can be classified under different categories when employed in different tasks. Let's consider how a toddler groups ground elements when playing with a brick set. She/he arranges elemental objects (bricks) to construct for instance a house, and thus, some bricks become the walls, others the roof, while other bricks make up the door of the house (see Fig. 4.1). However, she/he can physically (and also *mentally*) group the same objects in a different way (assigning them other, different concepts) to devise, for example, a dinosaur, in which now, some of the bricks that constituted the house walls may turn into a threatening head.

Based on these ideas, in this chapter we propose a particular multi-hierarchical symbolic model (*Multi-AH-Graph*), which corresponds to a subset of the *AGraph* category presented in Chap. 3, devoted to cope with multiple robot operations. Notice that each robot operation may require different arrangements of the symbolic data. Thus, we utilize a different hierarchy of our

**Fig. 4.1.** Different arrangements from a same set of ground pieces

multi-hierarchical model to solve each particular robot task. Here, we focus on three specific robot tasks: (1) task-planning, (2) self-localization, and (3) human-communication. Note that any robotic agent intended to attend users within human environments, like our taxi-driver, should be able to efficiently accomplish, at least, these three operations. The symbolic model can be easily extended to consider others (i.e., manipulation, surveillance, etc.) by adding extra hierarchies. We also describe in the following a *translation mechanism* to relate abstract symbols from one hierarchy to another. This translation is especially useful to shift concepts used to solve a certain robot operation, i.e. planning, to concepts understandable by a human, and thus, improving robot-human communication.

## 4.2 A Multi-Hierarchical World Model
##     for a Mobile Robot

In a nutshell, a *Multi-AH-graph* can be considered as a finite portion of the
*AGraph* category, in which arrows represent graph abstractions, while vertexes
are graphs. We consider only those portions of *AGraph* in which there exists
a unique initial vertex (there are no edges that end on it) that becomes the
*ground level* of the multi-hierarchy and represents world information with the
greatest amount of detail. On the other hand, we assume that there is only
one terminal vertex (there are no edges that start from it), called the *universal
level* of the multi-hierarchy that represents the most abstracted information
of the world. All different paths (sequences of arrows) between the ground
level and the universal level are single hierarchies, that is, *AH-graphs*[1].

As commented in the introduction of this chapter, we propose that each
hierarchy of our model is especially suitable for solving a particular robot
task. In general, the larger the number of robot tasks, the larger the number
of hierarchies. However, when dealing with complex or large environments we
can solve different instantiations of the same task through several hierarchies.
For instance, we can improve path-planning in large-scale environments by
using different hierarchies which are selected depending on the origin and the
destination of the path to be found [50]. Therefore, the relation *number-of-
hierarchies ↔ number-of-tasks* is just a broad guideline.

Next, we describe the *Multi-AH-graph* of our mobile robot through the
description of its different hierarchies. We focus on a robot that is intended
to carry out efficiently three operations: task-planning, self-localization, and
human-robot communication.

### 4.2.1 Task-Planning Hierarchies

The goal of the task-planning hierarchies of the multi-hierarchical model is to
improve efficiency in the planning process, which is required for deliberative
agents.

In the previous chapter, only one hierarchy was considered to improve task-
planning. Such a hierarchy grouped spatial information using human concepts
like rooms, areas, buildings, etc., for a clearer description of our hierarchical
planning approach. However, in practice, such an arrangement of spatial information may not lead to the best planning results in terms of efficiency and/or
optimality. Thus, we can consider not only a single hierarchy, but to use multiple of them to better adapt the model with respect to different planning
problems. In addition, when more than one problem is to be solved, the use of
more than one hierarchy may allow the system to solve part of them simultaneously [50]. The availability of more than one task-planning hierarchy is also

---

[1] Formally, the abstraction relation forms a partial order in the set of graphs of a
Multi-AH-graph, with one minimal and one maximal element.

convenient for planning under different criteria of optimality (length of plans or computational efficiency, for example). That is, a hierarchy may adapt better to solve tasks with low computational cost, while other may be oriented to obtain optimal plans (both criteria are often incompatible and therefore rarely are met together with only one hierarchy). These and other results on the advantages of using multi-hierarchical graphs for planning (in particular for route-planning) have been stated elsewhere [49].

Task-planning hierarchies can be automatically constructed as will be explained in Chap. 5 (elsewhere [49], a simpler algorithm based on hill-climbing has been presented). Such an automatic construction is guided to optimize over time the tasks to be performed by the robot, and thus, certain abstract concepts constructed in this way may not correspond to understandable concepts by humans, being necessary a translation to those concepts that the user understands. Notice that in Chap. 3, the hierarchy used in the experiences was constructed imitating the human cognitive map [89], and thus, no translation was required since planning results already involved human spatial concepts. We will delve into a translation process between concepts from different hierarchies in Sect. 4.3.

### 4.2.2 Localization Hierarchy

*Self-localization* and *mapping* are unavoidable robot abilities when it is intended to operate within unknown or dynamic environments. Both of them should be carried simultaneously (known as *SLAM*, for Simultaneous Localization And Mapping) by registrating new sensor observations (*local maps*) against a cumulative *global map* of the environment. From this process, the position and orientation (pose) of the robot relative to a global reference frame is estimated, and a global map is updated with the new data (see [147] for a comprehensive SLAM survey).

Estimating the robot pose through SLAM approaches in large-scale environments has been largely treated in the robotic literature [18, 25, 41, 86, 96]. Most of them considers two views of the robot environment: one involving geometric information (local maps) and the other one considering topological information, including global maps as the result of fusing a number of local maps [96]. The main points in dividing the environment into areas is to reduce computation cost on large amount of data and also to reset the odometric error accumulated during navigation [39]: if only one big map had been used it would produce unacceptable localization errors for large environments and an unaffordable computational cost [85].

Following this trend, we have employed in our experiments a 2-level hierarchy, called the *Localization Hierarchy*, devoted both to estimate the robot pose in an accurate manner and to update geometrical information (mapping) during navigation and exploration. We rely on an automatic procedure to detect set of observations whose local maps exhibit certain resemblance, and thus, that can be fused to produce consistent global maps [14]. More

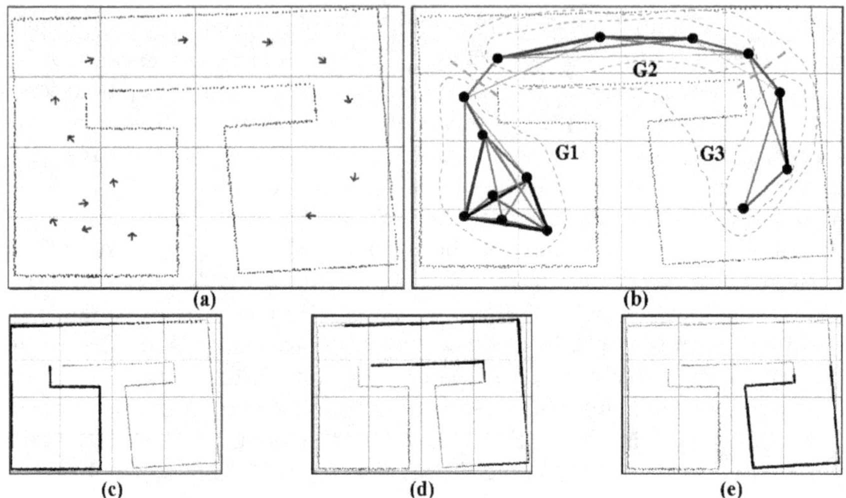


**Fig. 4.2.** Maps used in robot localization: (a) Global map automatically constructed during a robot navigation; (b) Automatic partition of the global map into submaps; (c)-(d) The three resultant local maps. Notice how the map partition process tends to divide the global map into open areas such as rooms and corridors, although this is not necessary

precisely, we use 2D laser scans to construct a topology in which each vertex (that corresponds to a robot pose) holds a point-based (local) map taken from such a location. Subsequently, the created topology is grouped taking into account a consistent measure (i.e., the overlapped area) for merging a set of local maps into a global map that represents a certain region. Figure 4.2 shows an example of how this technique works. Similar approaches to detect consistent regions in space can be found in [19] [24].

Thus, the localization hierarchy consists of only one hierarchical level (graph) built upon the ground level of the multi-hierarchy. Vertexes from the ground level represent local observations and hold geometric information (point-based local maps in our case) while vertexes from the first level, that represent regions of space, hold global maps resulting from fusing a set of local maps[2]. The first level is in turn abstracted to the universal level (which only contains one vertex representing the whole environment). The abstraction function for vertexes we use is defined upon a graph spectral partitioning algorithm as explained in [14].

Regions represented by vertexes from the first level of the localization hierarchy should facilitate the localization of the robot, i.e., by means of a coherent set of observations. Thus, each region corresponds to a group of nearby observations that share a sufficient amount of information. For example, the

---

[2] This creates a topology of space similar to the causal level of the Spatial Semantic Hierarchy [90].

observations acquired inside a room are usually abstracted to a different region than the ones acquired from an adjacent corridor, since localization within the room uses a different point-based map than localization within the corridor. This kind of regions bears a strong resemblance to that used by humans when navigating: we pay attention to different landmarks in different spatial areas. This provides the localization hierarchy with certain degree of similarity to the human cognitive map, although such requirement is not mandatory[3].

### 4.2.3 Cognitive Hierarchy

The following paragraphs describe how a hierarchy can be included within the multi-hierarchical symbolic model for interfacing with the *human cognitive map*. This type of hierarchy is necessary if the robot has some type of intelligent interaction with persons (like in the robotic taxi vignette). Although the previous two types of hierarchies can be automatically constructed (Chap. 5 details an evolutionary algorithm to construct planning hierarchies), due to its nature, the cognitive hierarchy should be generated with the human aid, so its length and shape may vary largely (see Fig. 4.3)[4].

For the case of spatial regions, vertexes of the hierarchical levels of our cognitive hierarchy will represent different groupings of distinctive places, defined by the human to directly match her/his cognitive map. Typically, local groups of distinctive places within a room are grouped into a supervertex that represents the room, groups of rooms and corridors are grouped into floors, floors are grouped into buildings, and so on. The main goal is to reproduce in the multi-hierarchical model the cognitive arrangement of the space made by the human who is going to communicate with the robot. The purpose of such a cognitive hierarchy is the improvement of the communication and understandability between the human and the vehicle. In this way, the person can request robot tasks using her/his own concepts (i.e., *go to the copier room*), at the same time that she/he can obtain robot information, for example results of task planning, by means of understandable terms (see Sect. 4.3). It is also possible to construct in this way more than one cognitive hierarchy, for communicating to more than one human.

We have used a quite simple construction process for the cognitive hierarchy (similar to the one presented in [87]). The user must select a number of vertexes from the common ground level (that represent observations) or from a certain level of the cognitive hierarchy (areas, regions, etc.) which will be abstracted to a vertex at the next higher level.

---

[3] For localization purposes, a large room can be split into two different regions, while the user may only identify a symbol for that room.

[4] This is because the cognitive hierarchy provides the particular viewpoint of each user, which may largely vary from others'. For example, an employee can group a set of distinctive places from the coffee room into a spatial concept labelled as "my rest room", while the person responsible for its cleanliness could group them, plus places from the near lavatories, under the label "the nightmare area".

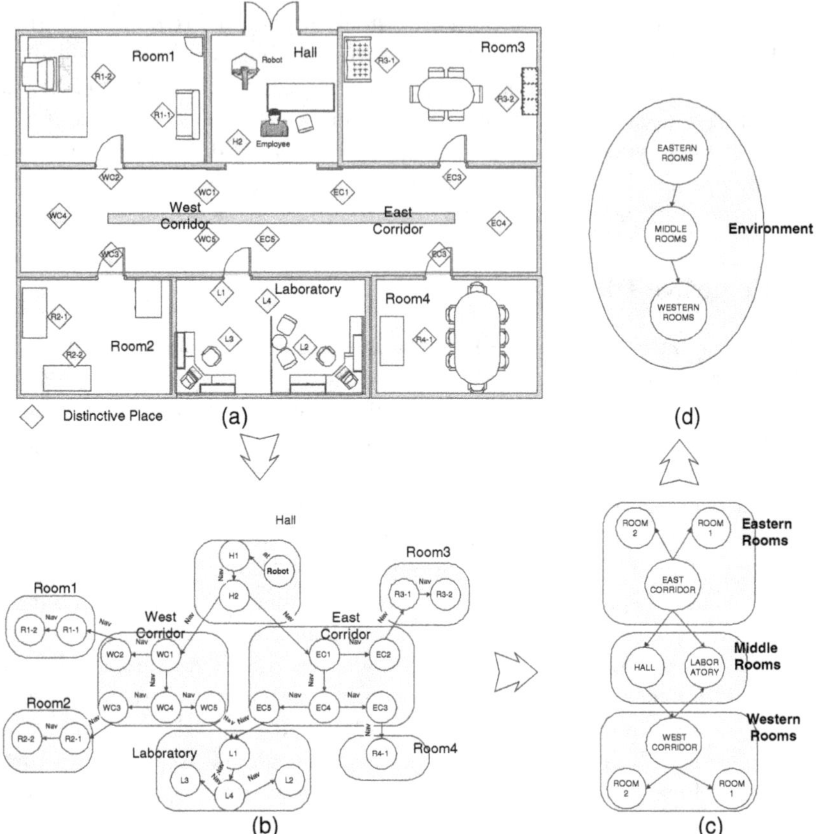


**Fig. 4.3.** Representation of an environment with different levels of abstraction: (a) The environment where a mobile robot may perform different operations; (b) Ground level: topology of distinctive places identified by exploring the environment. This will be the lowest hierarchical level of the Multi-AH-Graph; (c) Descriptive hierarchy for communications from/to human. This hierarchy represents the cognitive interface with the human driver

Here it gains relevance the way in which the ground symbols are acquired. As we will expose in detail in Chap. 5, acquiring symbolic information from sensorial raw data is an arduous problem not completely solved yet. A way of addressing it is the topology construction based on observations described before. Another approximation can be a process that automatically detects distinctive places, like entrances to rooms [24]. The result of such a process (vertexes representing those distinctive places) will be the starting point in the human-guided construction. The user could name ground symbols, for instance *"entrance door to my office"*, and group a number of them to create a supervertex with a special meaning for her/him, i.e. all these doors connect to the *"main corridor"*.

But as the reader may realize, the mechanism to automatically acquire symbolic information is not available yet (this is called the *anchoring* [28] problem). Thus, apart from the existence of an anchoring process to automatically acquire symbolic information, the user should also be enabled to provide ground symbolic information manually. For example, the user can guide the robot to a distinctive place that is of interest to him/her, though it could not be considered by the anchoring process, i.e. "in front of my desk". In this way a person can create a number of ground spatial symbols to represent the location of, for example, a cabinet, a window, or a pile of books, which may be jointly grouped with automatically acquired symbols, i.e. an observation, to make a supervertex that represents an entire room which could be labelled by the user as *"my office"*.

In any case, when vertexes at the ground level are (manually or automatically) created, geometrical information required for posterior robot operation, like maps, images, etc, can be automatically annotated. Edges can also be added to the model to indicate navigability (from the previous robot location to the current one) or position relations (a book is *on* the desk) [47].

In our experiences detailed in Chap. 6, the user of a robotic wheelchair can use voice commands to create a distinctive place for navigation ("I am interested in this place, its name is ⟨place-name⟩") or an abstraction of them ("The visited places up to now are grouped into a room called ⟨room-name⟩"). A representative example of a human-guided constructed cognitive hierarchy is depicted in Fig. 4.3.

## 4.3 The Utility of the Multi-Hierarchical Model for Human-Robot Interaction

As commented, a multi-hierarchical model like the one depicted in Fig. 4.4 can be used by an intelligent agent to efficiently perform different tasks in a hierarchical fashion. In that case, hierarchical approaches to solve certain operations, like task or route planning [60] [50], may provide abstract solutions before a concrete ground solution is achieved. This feature becomes very interesting when the robot is intended to operate with humans. Recall the taxi driver vignette: when the user asks the robotic driver to go to a particular place, it can start driving immediately (after it plans a route) or, on the other hand, it can report the user with relevant information about the planned route before it is executed. Although the first case is common when we catch a cab, the second option should also be desirable. Nevertheless, in this latter case, the user could be overwhelmed if the driver gives her/him highly specified information, enumerating streets and junctions, while it would be more useful to provide only abstract information at different levels of detail (about areas or districts to be traversed).

When the user is informed about the driver intentions, she/he can agree with the selected route to arrive to the destination, or in contrast she/he

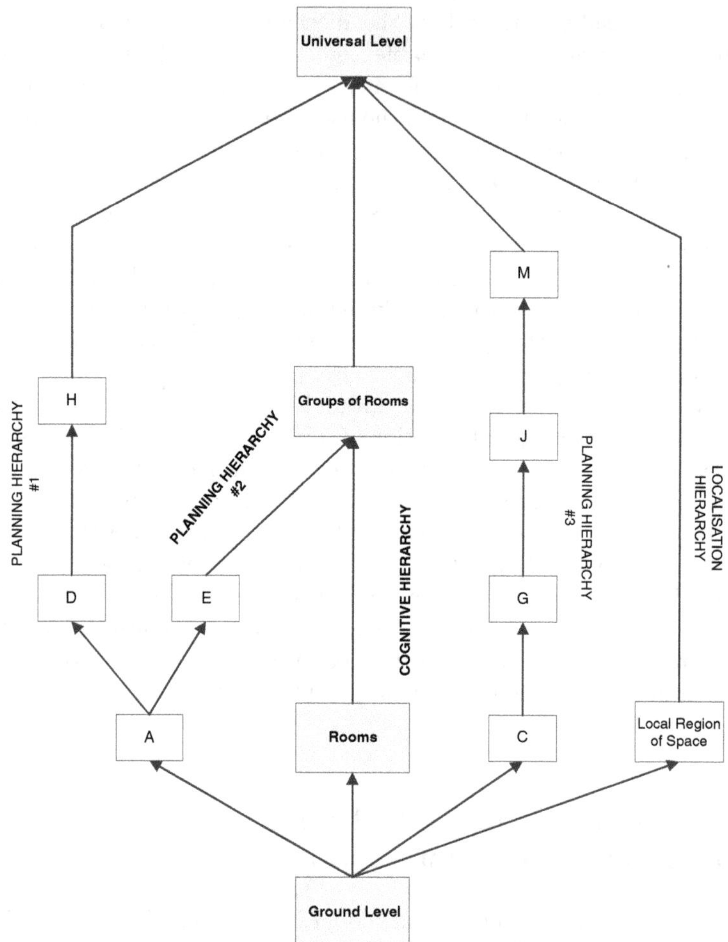



**Fig. 4.4.** An example of multihierarchy in which the cognitive hierarchy is highlighted. In this example we consider three different hierarchies for task-planning. Notice how hierarchical levels can be shared between different hierarchies

can advise an alternative one. In any case, abstract results from the planning process carried out by the driver should be communicated to the user in a proper way.

In this example, the taxi-driver planning process uses a particular hierarchy (possibly automatically constructed) which may largely differ from the cognitive one, and thus, abstract symbols from the former could not be understandable by a human. In order to report properly abstract information from one hierarchy, i.e. a *task-planning hierarchy*, to the user, a *symbol translation* to the cognitive hierarchy is required. Such a translation, that can be used to shift concepts between any pair of hierarchies, can be used to enable humans to interact with the robot task-planning process [58, 59]. The main advantages of interacting with a task planning process are:

1. A user can command the robot a task that involves abstract concepts from the cognitive hierarchy while the robot can carry out the task on a different, more efficient one.
2. The user can reject or accept the proposed abstract plan that solves a certain goal. Also, particular parts of the abstract plan can be accepted or rejected.
3. The user can ask the robot for more details about a given abstract solution.

Next, a general *inter-hierarchy translation process* that shifts concepts between hierarchies is detailed. Then, we describe an example of interactive task planning which involves symbol translations between the task-planning and the cognitive hierarchy described in previous sections.

### 4.3.1 The Inter-Hierarchy Translation Process

The *Inter-Hierarchy translation process* is aimed to translate symbols between any pair of hierarchies of the multi-hierarchical model. Broadly speaking, our translation mechanism consists of *refining*[5] a certain symbol from a hierarchy down to the common ground level of the model[6] and then, *abstracting* the resultant ground symbols along the destination hierarchy. However, while the abstraction of a given vertex (supervertex) of a graph in *A Graph* is unique, its refinement yields a set of subvertexes. Notice that the translation process may not produce an unique symbol at the target hierarchy, since when refining a vertex, a variety of subvertexes are obtained which are possibly abstracted to different concepts in the target hierarchy.

Let's consider an example in which two hierarchies represent different spatial arrangement of the streets of a city to illustrate the translation process. One hierarchy, the "Zip Codes Hierarchy" ($H_1$), groups streets following the administrative division for the post service, and the other one, the "Police Station Hierarchy" ($H_2$), groups streets covered by a given police station. The formalization of both hierarchies (see Fig. 4.5) is as follows (for simplicity we have not considered edges since they are not involved in this translation process).

Let the graphs (hierarchical levels) $G_0$, $G_1$, $G_2$, and $G_3$ be defined as:
$$V^{(G_0)} = \{s1, s2, s3, s4, s5, s6\}$$
$$V^{(G_1)} = \{Z1, Z2\}$$

---

[5] The proposed translation process can be successfully applied to Multi-AH-graphs that are subsets of the *CVA Graph* category, that is, those multi-hierarchies for which the refinement and abstraction functions are defined for any graph and are complete and covered.

[6] Although it is not a restriction of the *CVA Graph* category, we assume that all hierarchies of the Multi-AH-graph share a common ground level.

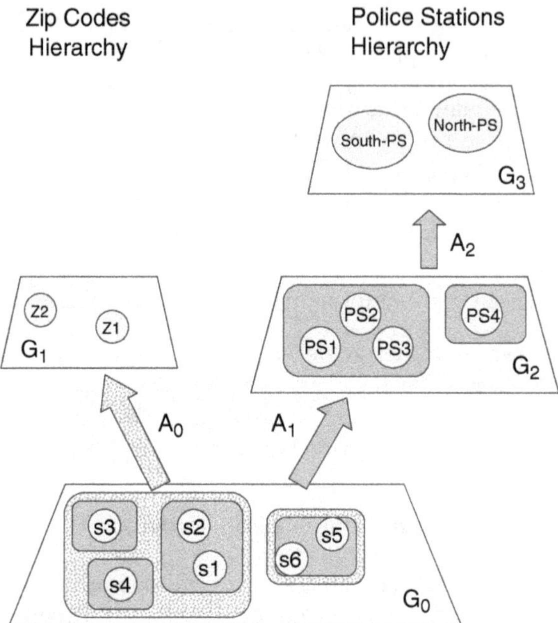


**Fig. 4.5.** Symbol translation between two hierarchies. A simple model with two hierarchies that groups symbols from the ground level (streets) with respect to its zip code (left) and the nearest police station (right). Dotted regions remark clusters of the zip code hierarchy while solid areas show clusters of the other hierarchy. For clarity sake, edges have been obviated in this example

$$V^{(G_2)}=\{PS1,PS2,PS3,PS4\}$$
$$V^{(G_3)}=\{South\text{-}PS,North\text{-}PS\}$$

and let abstractions on these graphs defined as:

$$A_0=(G_0,G_1,\nu_0,\emptyset)$$
$$A_1=(G_0,G_2,\nu_1,\emptyset)$$
$$A_2=(G_2,G_3,\nu_2,\emptyset)$$

where

$$\nu_0(s1) = Z1;\ \nu_0(s2) = Z1;\ \nu_0(s3) = Z1;\ \nu_0(s4) = Z1;\ \nu_0(s5) = Z2$$
$$\nu_0(s6) = Z2\ \nu_1(s1) = PS1;\ \nu_1(s2) = PS1;\ \nu_1(s3) = PS2;\ \nu_1(s4) = PS3;$$
$$\nu_1(s5) = PS4;\ \nu_1(s6) = PS4$$
$$\nu_2(PS1) = SouthPS;\ \nu_2(PS2) = SouthPS;\ \nu_2(PS3) = SouthPS;$$
$$\nu_2(PS4) = North$$

Hierarchy $H_1$ is a two-level hierarchy defined by abstraction $A_0$, while hierarchy $H_2$ is a three-level hierarchy defined by the composition chain $A_1 \diamond A_2$.

In this example the translation of a certain concept, let say $Z1$, from the "Zip Codes Hierarchy" into the other hierarchy will provide us information about the police station(s) that cover all streets grouped under the $Z1$ zip code. Such a translation starts with the refinement of $Z1$, yielding the set of ground symbols (streets and junctions) belonging to the same administrative area:

$$\nu_0^{-1}(Z1) = \{s1,\ s2,\ ,s3,\ s4\}$$

Since the ground level is common to both hierarchies[7], subvertexes of $Z1$ can be abstracted now through the target hierarchy, yielding the set of police stations that offer service to all streets with zip code $Z1$:

$$\nu_1(s1) = PS1,\ \nu_1(s2) = PS1,\ \nu_1(s3) = PS2,\ and\ \nu_1(s4) = PS3$$

That is, the translation of the symbol $Z1$ from the "Zip Codes Hierarchy" to the first level of the "Police Station Hierarchy" is the set $\{PS1, PS2, PS3\}$, which in its turn can be abstracted again to the symbol $SouthPS$.

Observe that in this example the translation of the symbol $Z1$ produces three different symbols at the first level of the target hierarchy, but a unique symbol at the second level. Thus, the ambiguity caused by the translation process can be solved by moving up at the target hierarchy, at the cost of loosing information. Next section details a human interactive task-planning approach that uses this inter-hierarchy translation process.

### 4.3.2 Interactive Task-Planning

Apart from the benefits for user communication, the user participation into the planning process may also improve its efficiency. In the hierarchical planning scheme explained in Chap. 3, task planning efficiency was achieved by detecting and removing unnecessary details of the robot environment. Now, an additional mechanism to discard useless information when planning can be provided by user interaction: when the user rejects or accepts an abstract plan (or part of it), she/he may discard world information, probably implying a further simplification in task planning. The benefits of this appear clearer if the robot deals with a very large environment like a real building composed of several floors, elevators, and hundred of rooms and corridors. In this case the reduction of the amount of data may become extremely important in order to plan efficiently robot's tasks.

We propose here an interactive, hierarchical task-planning approach based on the scheme described in the previous section, that can be split into two phases (see Fig. 4.6).

1. The first stage consists of translating the user task request, which can be specified at any level of the *cognitive hierarchy*, to the *task-planning hierarchy*. It is done by refining the human concepts involved in the task until the shared ground level (see Fig. 4.5). To do this, a particular refinement

---

[7] In fact, it is not necessary to refine symbols down to the ground level, but only to the first level that is common to both hierarchies.

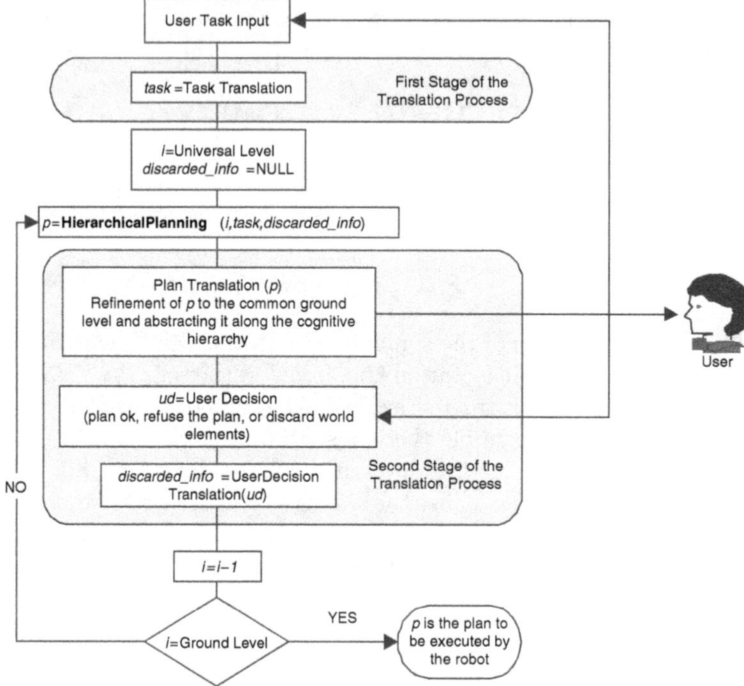


**Fig. 4.6.** The Translation Process for interactive task-planning. In the interactive planning process the user is well-informed about the sequence of abstract plans obtained by the hierarchical planner, allowing her/him to accept or reject plans as well as to discard non-desirable elements of the environment

of the concepts involved in the task is chosen[8]. For example, if the user request is "go to Laboratory", the human symbol (*Laboratory*) involved in this task can be moved down to the ground level, choosing as refined symbol $L1$, that represents the Lab entrance. Thus, the user task turns into "go to Laboratory entrance". Once the user task is specified at the ground common level, it can be solved through the planning hierarchy, choosing the appropriate hierarchy among the ones that are available.

2. The second stage is carried out in the opposite direction. A plan (or abstract plan) produced by the hierarchical planning process (HPWA-1 for example) should be reported to the user involving only concepts of the cognitive hierarchy. Moreover, the human decision about the produced plan (i.e. to reject or accept it) may involve a backwards translation to the planning hierarchy as shown further on.

The translation process for our interactive task-planning approach uses two mechanisms for refining/abstracting plans through the hierarchies of the

---

[8] This is similar to the action refinement procedure carried out in HPWA-2, described in Sect. 3.4.

model. Informally, refining and abstracting plans within a hierarchy consists of refining/abstracting the sequence of operators involved in the plan. Thus, abstracting a plan produces another plan with a lower level of detail, since it contains more general (abstracted) concepts. In contrast, refining a plan yields a set of more detailed plans covering all possible combination of the refinements for the parameters of the given plan. These functions, *RefPlan* and *AbsPlan*, can be formalized through the extended refinement/abstraction function for planning states given in Chap. 3 (Definitions 3.2.13 and 3.2.14) since a plan can be considered as a planning state[9].

Figure 4.3 depicts a scheme of part of the environment used in our experiments. Upon the ground level (that represents the robot environment with the maximum amount of detail) the cognitive hierarchy establishes a cognitive interface with the human [47]. Such a hierarchy enables the robot to manage human concepts for enhancing human-robot communication and interaction. The other one, the task-planning hierarchy, arranges properly the world elements with the goal of improving the hierarchical task-planning process. Figure 4.7 shows the multi-hierarchical model for that environment, comprising one cognitive hierarchy and one planning hierarchy. The former is manually constructed (refer to a previous work in [47]), while the latter is automatically constructed as described in Chap. 5.

In this scenario, let us consider the following robot application. An employee, at the entrance of the office building, is in charge of receiving and distributing mails to the rest of employees. To facilitate his work, a servant robot can carry objects within the office building, so he has only to give the proper envelope to the robot and select the destination, i.e. "go to the laboratory". Notice that the user specifies the task using the cognitive hierarchy and that she/he can choose the most convenient level of detail, in this example level 1 (see Fig. 4.7).

As previously commented, the first stage of the translation process consists of shifting the human concepts involved in the requested task into concepts of the ground level. Since the task *go to the laboratory* is not completely detailed (it is an *abstract task*) the robot has not enough knowledge about the particular distinctive place within the *laboratory* where it must go. This first translation phase is then solved by simply choosing a concept (vertex), for example the *laboratory entrance*, represented by the *L1* vertex at the ground level, embraced by the more abstract *laboratory* concept. Such a selection can be carried out following a certain criteria, i.e., the most connected one, a border vertex, randomly, etc. Refining the abstract task in this manner may cause that a feasible plan could not be found during the task planning process. In this case two possibilities are available: choose another subvertex or ask the

---

[9] More precisely, a plan is an ordered sequence of logical predicates, that is, it can be considered as a planning state in which its predicates are sorted. Therefore, when applying functions *EAbs* and *ERef* to abstract/refine plans, the order of the resultant plans must be kept.

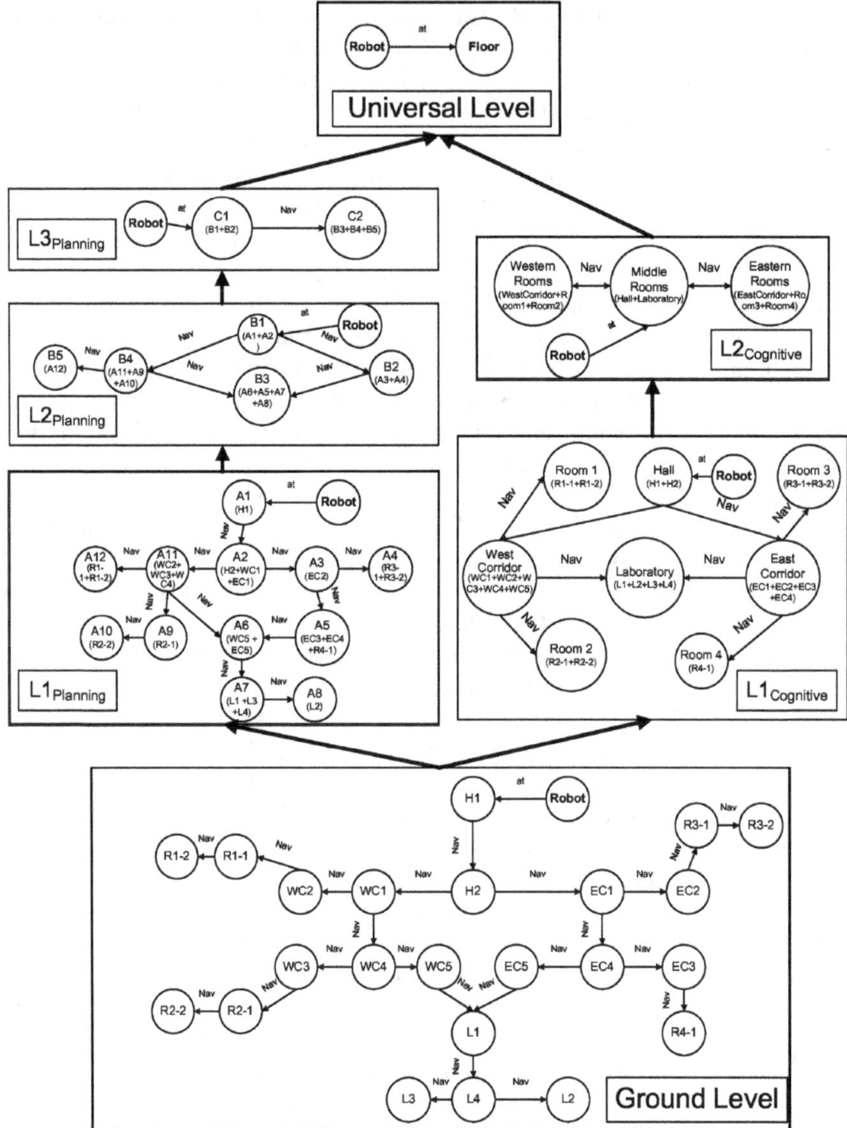


**Fig. 4.7.** A multi-hierarchical model of an office scenario. For simplicity, this Multi-AH-graph only consists of two hierarchies (planning -left branch- and cognitive -right branch-) which share the ground and universal levels. Notice that both hierarchies can contain different number of hierarchical levels. Labels of vertexes (in brackets) indicate the subvertexes that they embrace. Also note that the vertex that represents the robot is present at all levels to make planning possible (recall Chap. 3)

user for a more detailed specification for the task, i.e.: *go to the Peters' table in the laboratory.*

Once an adequate specification of the requested task is obtained at the ground level, the hierarchical planner (HPWA-1 for instance) solves the task using the planning hierarchy. Hierarchical planning produces successively abstract plans at different levels of the task-planning hierarchy, which may involve symbols not directly understandable by the user. Such symbols should be reported to the human using concepts from the cognitive hierarchy. This is the second stage of the translation process, which will make possible the user interaction with the task planning process.

More precisely, following our example, the planner produces the first abstract plan at level $L3$ from the planning hierarchy[10] which is {(GO C1 C2)}. This abstract plan must be translated into human concepts before reporting it to the user, since it entails two abstract concepts $C1$ and $C2$, produced by an automatic labeller process, which do not belong to the cognitive hierarchy.

Such a translation of the obtained plan at level $L3$ is achieved by refining and abstracting successively the parameters of the abstract plans through functions $RefPlan_{A_s}$ and $AbsPlan_{A_s}$. Thus, the refinement of the parameters of the plan {(GO C1 C2)}, down to the level $L2$ of the planning hierarchy, yields the list of plans (see Fig. 4.7):

$$\{(GO\ B1\ B3)\},\ \{(GO\ B1\ B4)\},\ \{(GO\ B1\ B5)\},\ \{(GO\ B2\ B3)\},\ \{(GO\ B2\ B4)\},\ and\ \{(GO\ B2\ B5)\}$$

that can be rewritten as a *plan schema*:

$$\{(GO\ \{B1,B2\}\ \{B3,B4,B5\})\}$$

By successively refining the abstract plans down to the ground level of the multihierarchy, a set of plans that only involve parameters which represent distinctive places are generated. Such plans are then abstracted up through the cognitive hierarchy through the $RefPlan_{A_s}$ function. In our example, this plan abstraction yields the set of plans shown in Table 4.8.

| | |
|---|---|
| Level 1 | (GO {Hall, East-Corridor, West-Corridor, Room3} {Laboratory, East and West Corridors, and Rooms 1,2, and 4}) |
| Level 2 | (GO {West, Middle, East} {West, Middle, East}) |
| Level 3 | N/A |

**Fig. 4.8.** Possible translations of the parameters of the abstract plan from level $L3$ of the planning hierarchy to the cognitive hierarchy. Notice how the plan ambiguity increases when it is translated to higher levels of the cognitive hierarchy

---

[10] The resultant plan at the universal level is trivial.

Once the plans are translated into the cognitive hierarchy, the user can proceed in the following ways:

1. *Inquiring a more detailed plan.* As shown before, the translation of an abstract plan may not provide enough information to the user. In these cases the user can request more information in two different ways. On the one hand she/he can ask the robot for a translation of the same plan using more detailed concepts of the cognitive hierarchy. Thus, the user obtains a more detailed information about the plan, but increases the plan ambiguity (please, see Table 4.8). On the other hand, the user can ask the planning process for planning a new solution at a lower level of the planning hierarchy. The obtained plan will involve more detailed concepts which can be translated into the cognitive hierarchy again, thus reducing ambiguity.

2. *Rejecting part of a plan.* Observe that even when the provided plan does not reveal enough information, the user can interact productively with the planning process, i.e. by rejecting certain spatial symbols. In our example when the robot provides the user with the translated plan using the level $L1$ from the cognitive hierarchy, she/he may require the robot to avoid the *West-Corridor* and *Room2* regions, since they are, for example, usually crowded. Such discarded symbols are translated again into the planning hierarchy (through refinement/abstraction functions), reporting to the hierarchical planner that symbols $B4$ at level $L2$ must be discarded. Thus, HPWA plans now at level $L2$ of the planning hierarchy without considering such a symbol, producing the plan {(GO B1 B2), (GO B2 B3)}, which is reported to the user as {(GO {Hall, East-Corridor} {East-Corridor, Room3}, (GO {East-Corridor, Room3} {East-Corridor, Laboratory})}.
Notice that all references to the "forbidden" symbols (those discarded by the user) and their subvertexes, have been eliminated.

3. *Suggesting an abstract plan.* Due to the ambiguity involved in the translation process, the user may be informed about a set of different possibilities to solve a plan. She/he can select one out of the offered solutions based on her/his knowledge of the environment. For instance, following the previous example, the user can suggest the abstract plan: {(GO Hall East-Corridor), (GO East-Corridor Laboratory)} since she/he knows that it is not necessary to consider *Room3* to arrive to the *Laboratory*. Thus, through the solution pointed out by the human, the planner can solve the task at the ground level considering only those symbols embraced by the ones suggested by her/him. In this example, the final plan at the ground level is (see Fig. 4.7):

$$\{(GO\ H1\ H2), (GO\ H2\ EC1), (GO\ EC1\ EC4), (GO\ EC4\ EC5),$$
$$(GO\ EC5\ L1)\}$$

## 4.4 Conclusions

This chapter has studied the use of different hierarchies upon a common ground level, that is, the use of a multi-hierarchical model, to improve several robot tasks at a time, including human-robot interaction. In particular, we have focused here on a translation process that enables the robot to communicate its intentions (the result of its planning process) to the user under such a multi-hierarchical fashion.

With this, we have assumed that the robot possesses a hierarchical arrangement of information to efficiently plan its tasks, whose results are communicated in a human-like manner to the user. What remains to "close the loop", that is, to permit the robot to autonomously and efficiently perform within a large environment, is to devise a mechanism to create and tune over time its multi-hierarchical and symbolic model. This can be carried out by methods like the one presented in the next chapter.

# 5

# Automatic Learning of Hierarchical Models

*Intelligence is the totality of mental processes involved in adapting to the environment.*
Alfred Binet, psychologist (1857–1911)

*One thousand days to learn; ten thousand days to refine.*
Japanese proverb

These two quotes summarize the key ideas of this chapter. On the one hand, it is common to relate the *intelligence* of an agent to its ability for *adaptation*. Informally, we usually consider as intelligent beings those that can face new problems or situations, that is, those that can adapt themselves to changes.

On the other hand, it is clear that efficiency is largely tight to the usual meaning of intelligence: it is common to consider more intelligent those beings that are able to solve problems (tasks) with less consumption of resources (time, energy, etc.). In this sense, an agent adapting to its environment should not only learn new information, but it should also arrange it (refine in the Japanese proverb) properly for an efficient use in the future.

In previous chapters, we have exposed mechanisms that permit agents to perform efficiently its tasks through a (hand-made) multi-hierarchical model. In this chapter we go into the automatic creation and optimization of planning hierarchies within that multi-hierarchical model of the environment[1]. In the following, a framework, called *ELVIRA*, is described. It is able to acquire and to arrange hierarchically symbols that represent physical entities like distinctive places or open spaces. Such an arrangement is tuned over time with respect to both changes in the environment and changes in the operational needs of the robot (tasks to be performed). This way, the work carried out

---

[1] We only focus on the optimization of the planning hierarchies since task-planning in large environments is highly affected by environmental changes, as well as it is one of the most complex tasks that a robot may carry out. However, the described approach can be extended to create/optimize other hierarchies of the model.

C. Galindo et al.: *A Multi-Hierarchical Symbolic Model of the Environment for Improving Mobile Robot Operation*, Studies in Computational Intelligence (SCI) **68**, 85–114 (2007)
`www.springerlink.com`                                © Springer-Verlag Berlin Heidelberg 2007

by ELVIRA is aimed not only to represent correctly symbolic information stemmed from the environment, but also to improve over time the robot efficiency in task-planning.


## 5.1 Introduction

The use of a symbolic representation of the environment becomes crucial in robotic applications where a situated robot is intended to operate deliberatively. In the literature there have been many types of environmental representations for robot operation (geometric [8, 66], probabilistic [147, 149], topological [26, 91], hybrids [120, 148], etc.). As commented in previous chapters, we are concerned with symbolic ones, in the sense that for deliberation, the information stored by the robot must represent the world through *qualitative* symbols (concepts) rather than by quantitative data acquired by sensors.

Such a symbolic representation permits the robot to intelligently deal with the environment, for example for planning tasks as explained in Chap. 3. However, maintaining a symbolic representation of the environment involves several important problems, some of then not completely solved or well understood yet. In this chapter we deal with three of them: (i) maintaining the coherence between the real world and the internal symbols, (ii) processing efficiently large amounts of information, which is an important problem in robots that work in real environments, and (iii) optimizing the model, that is, selecting among all the possible coherent set of symbols that one could use for modeling a given environment, the most appropriate for operating as efficiently as possible. Problem (i) is tightly related to the psychological *symbol grounding problem* [69], that deals with how humans categorize the world and their abstract ideas into concepts. In the mobile robotic arena, this issue has been simplified recently through the so-called *anchoring* [17, 28], although in other communities it is dealt with implicitly, such as in SLAM [67, 147] or in visual tracking [5, 75, 134]. Anchoring means to connect symbols of the internal world model of an agent to the sensory information acquired from physical objects. These connections can be reified in data structures called *anchors*, which contain estimations of some attributes of the objects. In that sense, an anchor is in fact a model of the corresponding physical object, and thus it can be used by the agent as a substitute for it, for instance in planning or when the real object is not visible. Notice that anchoring is not a static process, but it must update continuously the symbolic information (deleting old symbols, changing current ones' attributes, adding new symbols) by updating the anchors.

The problem of processing efficiently large amounts of information (problem (ii)) arises when real environments with a potentially huge amount of perceptual, quantitative information, are considered. In that case, the symbolic representation may also be large, and therefore it must possess a suitable structuring mechanism for enabling future accesses in the most efficient way. A common approach to that problem is the use of *abstraction*, which establishes different, hierarchically-arranged views of the world as we have seen in previous chapters. Some examples of efficient processing achieved through the use of abstraction can be found in [49, 50, 56]. It has been demonstrated that the use of abstraction can reduce the cost of processing information significantly, and even can make some exponential problems tractable.

Finally, optimizing the model (problem (iii)) arises since the robot does not carry out the same tasks all the time nor in the same environment. It should adapt to different tasks over its working life. Also, changes in the environment may lead to reduce the suitability of the current internal representation for task planning and execution, and thus require a better structure of symbols. Hence, there is a need for some procedure that tunes the symbolic representation dynamically, optimizing it with respect to the current knowledge of the tasks the robot has to plan and execute. This optimization procedure can be appropriately addressed by heuristic optimization techniques, i.e. *evolutionary algorithms* [9], as we will show later.

In this chapter we mainly focus on problems (ii) and (iii) although we also provide a framework where issue (i) can fit well. We propose an evolutionary algorithm adapted to be executed as an any-time algorithm (that is, the longer the algorithm is executed, the better the constructed symbolic representation, but there is always a correct structure available for the robot). For dealing with efficient structuring of the model, we use our *AH-graph* model which has demonstrated its suitability for improving robot operation within large environments in previous works [49, 56]. To prove the suitability of our approach within a real robotic application, our method has been integrated into a general framework in which the symbol grounding problem (i) fits. For coping with automatic anchoring, we use techniques for extracting topological maps from grid maps [24, 43], and also image processing techniques for anchoring objects [60].

In the literature, works can be found that endow mobile robots with a symbolic representation of the environment, most of them using topological representations [18, 145, 152, 156], and often for path-planning. They cope with the automatic construction of a symbolic representation of the robot space (topology), and thus, they approach the first commented problem when anchoring distinctive places from the robot environment to symbols in the topology. Some of them also apply this topological (or in, general hybrid) techniques to large environments by hierarchically arranging symbolic (and also some geometric) information [90, 96, 152]. However, not much attention has been paid to the optimization of that internal symbolic structure during the robot operating life. As Wah stated in 1989: "despite a great deal of effort devoted to research in knowledge representation, very little scientific theory is available to either guide the selection of an appropriate representation scheme for a given application or transform one representation into a more efficient one" [155]. Nowadays, such an affirmation is still applicable and only a few works aim to organize symbolic information [144, 146], but without pursuing the optimality of robot operation within dynamic environments. This chapter focuses on this less explored direction.

The framework presented here, that we call *ELVIRA*, has been fully implemented and tested through real and simulated experiments, demonstrating its feasibility for the automatic creation and optimization of large symbolic representations of the robot environment that are used for efficient task planning and execution [57].

Next section gives a general view of the ELVIRA framework. Following sections are devoted to go into its two main functions: Sect. 5.3 focuses on the creation of ground symbols that make up the model, and Sect. 5.4 delves into the automatic arrangement of such symbols into an efficient hierarchy for task planning. In Sect. 5.5 some discussion and experimental results are presented.

## 5.2 The ELVIRA Framework

The ELVIRA framework enables a mobile robot to solve correctly and efficiently its tasks within a dynamic and large environment by automatically creating an anchored symbolic representation, which is adapted over time with respect to changes in both environmental information and the set of tasks that the robot must perform. It has been fully implemented as a computational system (Fig. 5.1 shows the general scheme) with two inputs: the task to be performed by the robot at each time, and the environmental information gathered by its sensors, and two outputs: a plan that solves the requested task as best as possible given the knowledge acquired up to the moment, and a symbolic structure anchored to the environmental information and optimized for the set of tasks that the robot has dealt with.

We use sensorial information to create and maintain anchored symbols (rooms, distinctive places for navigation, and objects for manipulation) arranged in a planning hierarchy of the multi-hierarchical model of the environment. At the lowest level of the multi-hierarchy, vertexes represent distinctive places or objects from the environment, while edges represent relations between them, i.e. navigability between places or location of objects with respect to others. Upon this common ground level, which varies over time to capture the dynamics of the environment, a planning hierarchy is automatically constructed and maintained to improve the robot operations.

As shown in Fig. 5.1, our framework carries out the automatic creation and optimization of the robot's world representation through three processes: the *Anchoring Agent*, the *Hierarchy Optimizer*, and the *Hierarchical Task Planner*, which we consider periodically executed with periods denoted in Fig. 5.1 as $\tau_{anchoring}$, $\tau_{optimizer}$, and $\tau_{tasks}$. That is, every $\tau_{anchoring}$ time-units the ground level of the symbolic structure is updated with the information collected by the anchoring process; every $\tau_{optimizer}$ the optimization process works to improve the internal hierarchical representation; and every $\tau_{tasks}$ a

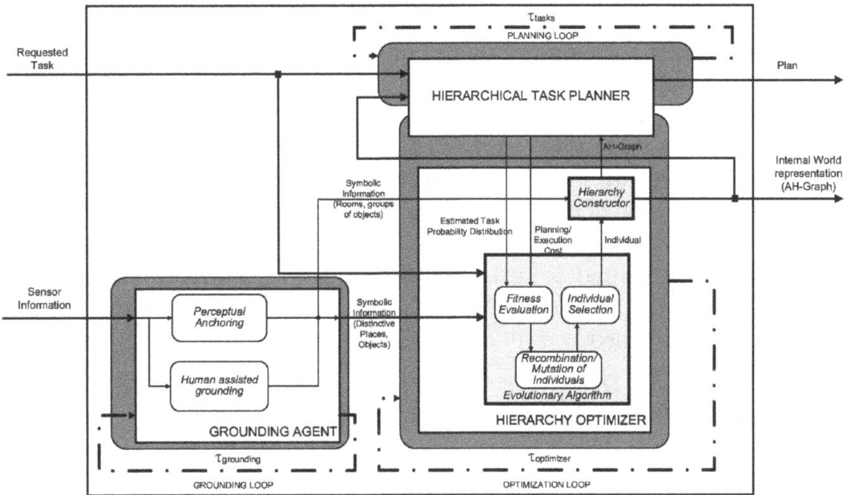


**Fig. 5.1.** A general view of the framework ELVIRA. It is fed with both the information gathered by the robot sensors and the requested task. It yields the best known planning hierarchy adapted to the agent tasks and environment, and a resulting plan for the currently requested task. ELVIRA includes three internal loops -anchoring, planning, and optimization- (marked in the figure by dark-gray shaded regions) that run concurrently

general task (not only navigation) is required to be planned and executed by the robot[2]. Notice that this period-based scheme, which is very useful for evaluation purposes, can be changed into an event-based one with minor modifications.

In the following, the internal processes of ELVIRA regarding the creation/maintenance of the symbolic model (Sect. 5.3) and the planning hierarchy optimization (Sect. 5.4) are described in detail.

## 5.3 Model Creation/Maintenance

All deliberative systems that rely on a symbolic representation of the robot workspace must face the *symbol grounding problem* [69] which is not completely solve yet. This problem, largely treated in the scientific literature [28, 34, 136, 139], is related to the genesis of symbols (concepts) stemmed from reality. From an objectivist point of view, in the case of symbols that represent physical objects, they can only be the result of many observations plus a mental process able to extract similarities from data. But how does such a mental process work? How could a robot imitate such a human ability?

---

[2] This sampling scheme implies that the internal world representation may be temporarily inconsistent with the environment for short intervals of time. This case is not considered in this book.

As the reader may realize, the symbol grounding problems becomes one of those wonderful mysteries of our brain: how does a young child make up a symbol for trees?, and how can she/he recognize the image, captured from her eyes, of a non previously seen tree?

These are questions that are out of our understanding now, but we are aware of their relevance for any system that accounts for intelligence. To cope with this issue, in this book we have largely simplified the robot acquisition of ground symbols by considering only distinctive places, open spaces, and simple objects (like coloured boxes), and providing two acquisition mechanisms that implement the Grounding Agent: a *human-assisted mechanism*, and an automatic process called *perceptual anchoring*, proposed previously in [28, 97].

### 5.3.1 Human-Assisted Symbolic Modeling

A naive solution to the symbol grounding problem is to include a human into the system to inform the robot when a new symbol must be created (or modified). Thus, we permit a human to become the *Grounding Agent* of ELVIRA. The robot, on its turns, links available sensory information, like odometry, maps, camera images, etc., to the symbols when they are created. That information can be used by the robot, for instance, for localization, navigation or for human-robot interaction as was explained in Chap. 4.

This solution exhibits great advantages in certain robotic applications. For instance, in assistant robotics, in which both the human and the robot work closely, it becomes useful that the symbolic world model of the robot mimics as much as possible the human one, and that is only possible if the human participates on the creation of it (i.e., in the cognitive hierarchy, explained in Chap. 4).

Let think, for example, of a robotic wheelchair that provides mobility to an elderly person, like the one presented in [64] and in Chap. 6. In this application symbols represent physical locations of the human/robot workspace that are of interest to the wheelchair user. But, how could a software algorithm decide what places are of interest to a particular user? As commented in Chap. 4, the human can indicate whether the current location of the vehicle (wheelchair), let's say *P1*, is a distinctive place, and therefore should be represented by a symbol. Symbols representing distinctive places are reified as vertexes at the ground level of the multi-hierarchical graph. Sensory information (percepts) is stored as annotations on vertexes, as well as geometrical information. During vehicle navigation (autonomously or manually guided) the user can inform about a new distinctive place, let's say *P2*, creating both a new vertex (for *P2*) and an edge (from *P1* to *P2*). This process can be repeated as many times as distinctive places the user selects in a certain area.

Following this procedure, we also enable the robot to manage more complex symbols. For example, when the user decides that the vehicle is leaving to what is considered by her/him a new area of the environment (for example, a new room), a symbol for that area is created at the first level of the hierarchy, as a supervertex that groups all the distinctive places sampled up to the moment.

This human-guided model creation is also a versatile process that permits the creation of cyclic topological maps when the vehicle returns to a distinctive place already visited, a problem not satisfactorily solved in the robotic literature yet.

### 5.3.2 Perceptual Anchoring

*Perceptual anchoring* or simply *anchoring* is a topic explored explicitly only in the recent robotics literature [17, 28, 55]. It can be thought of as a special case of the symbol grounding problem [69] where the emphasis is put on: (i) symbols that denote physical entities, and (ii) the maintenance of the connection between these symbols and the corresponding entities via perception while their properties change. Broadly speaking, anchoring consists of connecting internal symbols denoting objects to the dynamic flow of sensor data that refer to the corresponding objects in the external world. Such a connection is achieved by using the so-called *predicate grounding relation*, that embodies the correspondence between predicates in the world model and measurable values from sensor information. For instance, the symbol `room-22` that fits the predicates 'room' and 'large' could be anchored to a local geometric map with certain parameters (shape and dimensions) that corresponds to that room.

The anchoring process, as defined in [28], relies on the notion of *anchors*: internal representations of physical objects that collect all the relevant properties needed for (re-)identifying them and for acting upon them.

The anchoring process comprises three procedures which are concerned with the creation and the maintenance of anchors:

- *Find.* This is the first step of the anchoring process, in which an anchor is created for a given object in the environment the first time that it is perceived by the robot, that is, when a certain piece of sensor data fulfills the requirements of a predicate grounding relation, e.g., size and color in an image, which identify a particular object.
- *Track.* During this phase a previously created anchor is continuously updated to account for changes in its properties, using a combination of prediction and new observations.
- *Reacquire.* This phase occurs when an anchored object is re-observed by the robot sensors after it was not seen for some time since. This may happen because the object was occluded, or because it was out of the current field of view of the sensors.

In general, the properties stored in the anchor can be used to estimate the current object's parameters (i.e., its position) for the purpose of controlling action, or to recognize the object when it reappears in the sensor's field of view.

## The Anchoring Agent

Within ELVIRA, the Anchoring Agent implements a simplified perceptual anchoring process that uses color images and range-scan data to create symbolic representations of objects and of places. The main characteristics of the Anchoring Agent are as follows.

### The Anchors

We only consider three classes of world's elements: *rooms*, *distinctive places* inside them, e.g., the center and the entrance of each room, and *boxes*. Anchors for places and rooms are grounded on sensor data collected by the sonar sensors as explained below; their properties include the position of the place (or center of the room), plus a set of geometric parameters (e.g., room size). Anchors for objects (boxes) are grounded on images collected by a color camera, and include as properties the position of the object, together with its color, shape, and size. All the anchors are linked to ground symbols in the AH-graph world model.

When the anchors are initially created, they are maintained in the robot's "local perceptual space", where their position is stored in robot's coordinates. Anchors in the local perceptual space are continuously updated according to the robot's perceptions and motion. When the robot leaves the current room and enters another one, the anchors in the local perceptual space are discarded, and they are replaced by the ones corresponding to the objects and places in the new room. However, the properties of the discarded anchors are not lost, but they are stored together with the associated symbol in the AH-graph world model.

### Anchor Creation

According to the nature of the world entity to be anchored to the symbolic structure, we distinguish two different mechanisms: (i) anchoring of simple objects and (ii) anchoring of topological entities.

For (i) we use the well-known *CamShift* algorithm [5] to detect areas in the camera images with a given distinctive color (red and green in our experiments). When a portion of image meets such a restriction, that is, it can be considered as a box, a new anchor is created for it, storing the position of the box with respect to the current position of the robot as a relevant parameter of the anchor. This information will serve to relate the created symbol that represents the box to the distinctive place where the robot is [60]. Other information included in the anchor is the color and size of the box, in order to distinguish it from other boxes in nearby positions.

Distinctive places as well as open spaces (mechanism (ii)) are anchored using the technique for topology-based mapping presented in [43], although

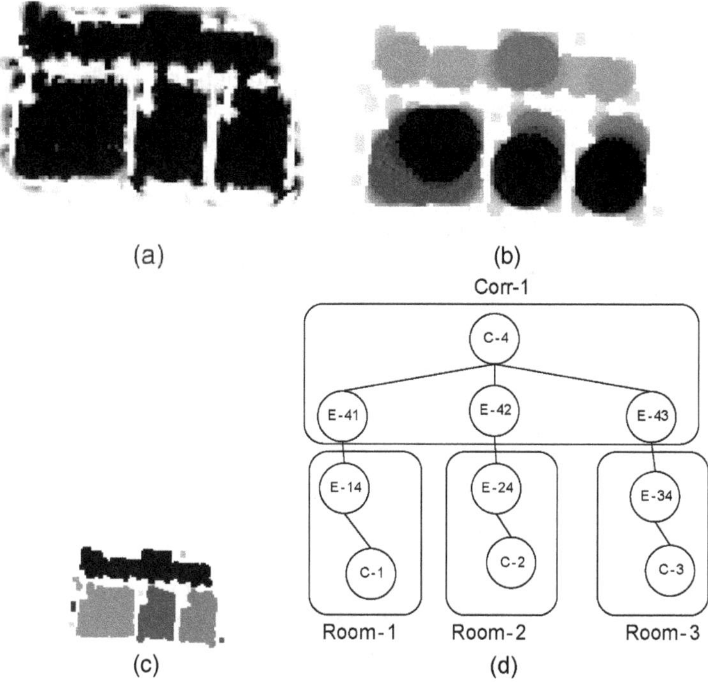



**Fig. 5.2.** (a) Original grid-map; (b) Fuzzy morphological opening; (c) Watershed segmentation; (d) Spatial hierarchy involved in this example. Distinctive places are marked with $Ci$ (the center of room $i$) and $Eij$ (the connection from room $i$ to room $j$)

other methods for topological map building could also be used ([26, 92, 90, 96, 148]). Our technique uses the data gathered from ultrasonic sensors to build an occupancy grid-map of the surroundings of the robot (see Fig. 5.2a). This grid-map can be seen as a gray-scale image where values of emptiness and occupancy correspond to gray-scale values. We then apply image processing techniques to segment the grid map into *large open spaces*, that are anchored as rooms.

Using the segmented grid map, the Anchoring Agent can also compute some room information like the dimension and shape of regions, as well as distinctive places of the topology such as the center and connection points between rooms. It should be noted that this technique relies on the assumption that the robot operates in indoor environments which have a room/corridor structure. Under this assumption, this technique has been shown to produce reliable maps [24].

*Anchor Maintenance*

Maintenance of anchors, in general, is made through the Track and Reacquire primitives by: (1) predicting how the properties in the anchor will change over time; (2) matching new perceptual information with these predictions; and (3) updating the properties accordingly. We perform in this paper a simple maintenance for anchors: the only time-varying property that we consider is the relative location of objects with respect to the robot. Thus, prediction coincides with self-localization of the robot, which can be solved through well-known methods ([67, 147]).

The maintenance of anchors stemmed from objects (boxes) implies searching in the model for previously anchored objects with similar parameters, that is, at the same location.[3] If there exists an anchored symbol in the model which is recognized as the one currently detected, its relative position is updated; otherwise, it is considered as a new object, and thus, a new anchor is created. (See [97] for more on perceptual management strategies.)

For topological anchors, new perceptual information is fused with the current map [42]. If the fusion is good, the robot updates its self-localization (or, from a dual point of view, it updates the relative position properties of the anchors). If not, the robot may create new anchors corresponding to the newly observed entities (rooms or open spaces).

It should be noted that anchoring is, in general, a very difficult problem, which has to cope with the great uncertainties stemming from the interpretation of perceptual data. A reliable treatment of anchoring would have to deal, among other things, with the problems of uncertainty in the perceived properties, ambiguity in data association, and multi-modal sensor fusion. (See, e.g., [23, 55, 97] for some work on these problems in the context of anchoring.) Since the focus on this paper is on the optimization of a symbolic model, however, we have preferred to use here a simple version of the anchoring process that solves ambiguous cases by simply selecting the best matching candidate according to simple metrics. As noted above, we have used probabilistic techniques for position update (through SLAM), but here too we only keep the most probable hypothesis. This simple approach was sufficient in our experiments, since we have used easily distinguishable objects (colored boxes) well spaced between them. In practice, we have also relied on a human supervisor to spot and avoid possible errors in the anchoring process. Clearly, a fully reliable autonomous robot would need to use more sophisticated techniques for vision, localization, and anchoring, but investigating this is out of the scope of this paper.

---

[3] Since the errors in estimating robot pose are around 4 cm (following the approach presented in [15]), and the expected errors in computing the relative position of objects with respect to the robot (around 15 cm), we have considered that two objects are the same when they are separated less than 20 cm.

## 5.4 Model Optimization

As commented at the beginning of this chapter, efficiency in task planning and execution is a relevant ingredient to define intelligence. Although the creation and maintenance of a coherent internal model with respect to the real world is an unavoidable issue for the correct planning/execution of tasks, the proper arrangement of such a model is also significant for the robot success. A vast collection of symbols arising from a real scenario arranged in an inappropriate manner may turn the execution of robot tasks into an impossible mission. What is the purpose of a highly coherent and updated model when it is not useful, for instance, for planning or reasoning? One can compare this situation to a library which entails a wide and valuable collection of books stored out of any order (see Fig. 5.3). Is that amount of valuable information useful when you have to spend days, weeks or even years to find what you are looking for?

As books in libraries are ordered following some criteria to facilitate the search, symbols managed by autonomous agents should also be arranged for the ease of their manipulation. The question is: what is the most convenient way to organize symbolic information? In the case of a library, books are indexed by genre, title, or author, since most of searches are based on these keys. Following this idea, symbols from the internal model of a mobile robot should be arranged in a certain way to improve over time the most frequent tasks performed by the robot. This is the key point of the model optimization part of the ELVIRA framework. As commented, ELVIRA considers symbols to represent distinctive places and simple objects, which must be arranged properly in order to improve the robot tasks. We focus on hierarchical structures since they are helpful in coping efficiently with the huge amount of information that a mobile robot may manage [56]. But the number of different hierarchies that can be constructed upon a set of symbols (as well as the different permutations of a set of books on a shelf) turns the model optimization problem into a hard problem for which no polynomial-time algorithm is known to exist.

**Fig. 5.3.** *"Order is never observed; it is disorder that attracts attention because it is awkward and intrusive"* (Eliphas Levi)

A feasible solution to hard problems is the use of evolutionary algorithms. In particular if they can be used as any-time processes, that is, the more they work, the better is the solution [61, 151], we find a good approach to our problem since in our case, finding the best hierarchy to solve the robot tasks (the most frequent ones) involves testing all possible candidates in the combinatorial space of hierarchies. The reader can find a vast literature about the use of evolutionary algorithms. Some examples are [52, 103, 113, 121], that use evolutionary processes for adapting the behavior of mobile robots, pose estimation, etc. Evolutionary approaches for graph clustering, similar to the model optimization process described in this chapter, can be found in [102, 125]. An analysis and in-depth classification of evolutionary computation can be found in [9].

Next section gives a general description of evolutionary algorithms, detailing our particular implementation.

### 5.4.1 Evolutionary Hierarchy Optimizer

*Evolutionary computation* is a biological inspired technique that simulates the natural evolution of species in which only those individuals best adapted to the environmental conditions survive. In nature, individuals of a population are affected not only by the environment conditions (climate, food abundance), but also by the rest of individuals. Individuals that perform better than others have a higher chance to survive, and thus, a larger opportunity to perpetuate their genetic material. This evolutive process ensures the improvement of individuals over generations.

Evolutionary computation, mimicking the natural evolution process, considers a set of individuals (*population*) that represent particular solutions to a problem to be solved, while the environmental conditions correspond to the evaluation of each solution (*individual*) with respect to a certain *objective function*. Such an objective function measures the individual adaptation to the particular problem we want to solve.

Broadly speaking, evolutionary algorithms are iterative processes aimed to find a good solution to a certain problem (see Fig. 5.4 for a general scheme). Typically, the process starts from an initial population generated at random, in which each individual provides a solution to the problem at hand. The evolution of individuals (solutions) is carried out through genetic-based operators: *recombination*, that simulates the cross between two individuals in which genetic information is interchanged, and *mutation*, that simulates a sporadic change in individuals' genotype.

At each iteration (also called *generation*), individuals are evaluated with respect to the objective function. The result yields the degree of adaptation of each individual to the environment, that is, to the problem. Those individuals with the best results pass to the next generation with a high probability of

```
generation=0;
P=Initialize();
Evaluate(P)
repeat
                P'=Variation (P)
                Evaluate(P')
                P=Select(P')
                generation=generation+1
until end condition
```

**Fig. 5.4.** Pseudocode of a general scheme for evolutionary algorithms. There are three relevant processes: (i) *Variation* which produces changes in a given population based on recombination and mutation operators; (ii) *Evaluate* which yields the adaptation of each individual to the problem to solve, and finally (iii) *Select* that decides which individuals deserve to pass to the next generation

recombination and/or mutation, while individuals with the worst results will not participate in the next evolution stage (they are eliminated).

When applying evolutionary techniques to solve a problem some considerations must be set. First of all, it must be defined the vector of parameters that encodes a solution to the problem at hand, that is, we have to define the genetic material of each individual, which is normally divided in parts called *chromosomes*. Next, depending on the particular encoding of the genetic material, recombination and mutation operators must be properly defined to permit the evolution over generations. And finally, the "evolution rules" must be set, that is, the process to select individuals that will undergo mutation and/or recombination and the process to select individuals that will be eliminated/maintained after each generation. We have chosen typical evolutionary techniques, such as the *roulette wheel* for selecting individuals and *elitism* for keeping the best individuals over generations. We recommend [9, 123] for a further description of these and other techniques used in the implementation of evolutionary algorithms.

In the model optimization carried out by the Hierarchy Optimizer within the ELVIRA framework, the search space is the set of all possible hierarchies that can be constructed upon a set of grounded vertexes (symbols). Thus, each individual of the genetic population represents a hierarchy of symbols. Therefore, we have to firstly give the set of parameters that univocally defines a hierarchy (a solution), the recombination/mutation operators that make possible the evolution of individuals over generations, and the objective function that guides the evolution process.

### 5.4.2 Population

The Hierarchy Optimizer starts from a random population of hierarchies that represent different valid possibilities of structuring the symbolic information of the environment. These individuals encode the minimum set of parameters, called *strategy parameters*, to generate the upper levels of AH-graphs upon a ground symbolic level that is maintained by the Anchoring Agent.

An individual of the population (an AH-graph) can be constructed by clustering algorithms that group nodes of a certain hierarchical level to produce the next higher level [49]. We have implemented a clustering algorithm (see the pseudo code in Fig. 5.5) that generates clusters given a set of seed nodes: initially, each cluster contains only one seed node, while the rest of non-seed nodes are iteratively added by following a deterministic order in the connectivity of the graph.

Since each hierarchy is a sequence of clusterizations starting at a ground graph (the ground level), we can define it as a set of *chromosomes*, where each one holds information to create a hierarchical level from a lower level through a set of seed nodes. Thus, an individual $i_a$ will represent a hierarchy with $k$ levels constructed upon the common ground level by encoding a vector of *k-1* chromosomes:

$$i_a = \{c_1, c_2, \ldots, c_{k-1}\} = \left\{ \left\{ n_1^1, \ldots, n_p^1 \right\}, \left\{ n_1^2, \ldots, n_m^2 \right\}, \ldots, \left\{ n_1^{k-1}, \ldots, n_r^{k-1} \right\} \right\}$$

$$(5.1)$$

```
PROCEDURE Clustering (Graph g, Seeds s):Clusterization cl
cl={}
FOR i = 1 to size(s)
                cluster = new cluster( s(i) )
                cl=cl + cluster
END
WHILE (UnclusteredNodes()==TRUE)
                FOR j= 1 to size(cl)
                        out=ConnectedNodes(cl(j))
                        FOR k=1 to size(out)
                                IF (Unclustered(out(k))
                                        cl(j)=cl(j) + out(k)
                                END
                        END
                END
END
RETURN cl
```

**Fig. 5.5.** Pseudo code of the implemented clustering algorithm. Initially it creates a set of clusters each one containing a single seed (a node from the given graph). These clusters grow up progressively including connected nodes, in a deterministic order, that do not belong to other clusters. The process finishes when all nodes are clustered. The function *ConnectedNodes* explores the outgoing arcs from a given node to yield the set of its neighbor nodes, while *Unclustered* returns true for a given node when it is not already included in a cluster and *UnclusteredNodes* returns true while nodes from the graph $g$ remain not assigned to a cluster. Notice that the algorithm must consider isolated nodes as unitary clusters in order to always end

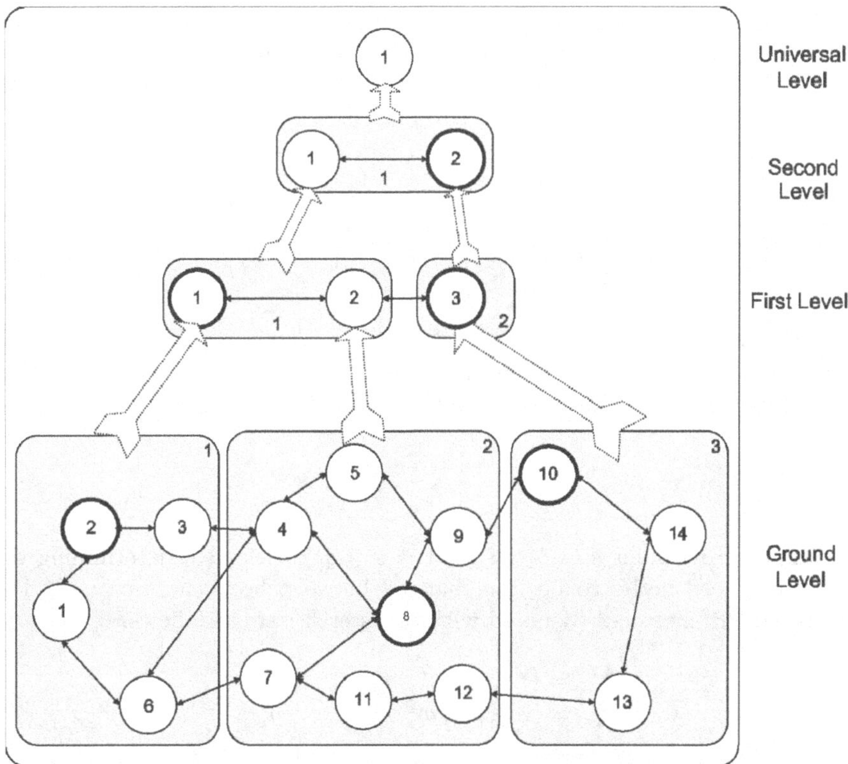

**Fig. 5.6.** Example of graph clustering. This is the hierarchy encoded by the individual ($i$={ *(2,8,10), (1,3), (2)*}) from a given ground level, after executing successively the clustering algorithm of Fig. 5.5. Seed nodes are indicated by thick circles

where $n_i^q$ is the unique identifier of the *i-th* seed taken from level $q$ that will produce a cluster at level $q + 1$. Seeds of the first chromosome refer to nodes of the ground level (that is, symbols acquired directly by the Anchoring Agent), while seeds from the other levels refer to clusters previously generated. Thus, for instance, $n_i^2 = j$ indicates that the *j-th* cluster generated by a seed node of the first chromosome is considered as the *i-th* seed for clustering the second hierarchical level. Figure 5.6 shows an example of the resultant hierarchy encoded by the 3-chromosome individual $i$={ *(2,8,10), (1,3), (2)*} upon a given ground graph.

### 5.4.3 Individual Recombination

In our hierarchical optimization approach, we combine two individuals (hierarchies) by interchanging part of their genetic information (clustering information), that is, individuals involved in recombination interchange part of their seed nodes. The result is that the original individuals are turned into two new ones (their offspring) containing a mix of the original genetic material.

Formally, our process for recombination is as follows. Given two individuals $ind_1$ and $ind_2$ made up of $k$ and $l$ chromosomes, respectively:

$$ind_1 = \{c_1^1, \ldots, c_k^1\} = \left\{ \left\{ n_1^1, \ldots, n_a^1 \right\}, \ldots, \left\{ n_1^i, \ldots, n_w^i \right\}, \ldots, \left\{ n_1^k, \ldots, n_c^k \right\} \right\}$$

$$ind_2 = \{c_1^2, \ldots, c_l^2\} = \left\{ \left\{ m_1^1, \ldots, m_d^1 \right\}, \ldots, \left\{ m_1^i, \ldots, m_q^i \right\}, \ldots, \left\{ m_1^l, \ldots, m_f^l \right\} \right\}$$

an $i$ chromosome index is selected at random, where $1 \leq i \leq k, 1 \leq i \leq l$. Let $c_i^1$ and $c_i^2$ be the selected chromosomes of individuals $ind_1$ and $ind_2$, that have $w$ and $q$ seed nodes respectively:

$$c_i^1 = \{n_1^i, \ldots, n_w^i\}$$
$$c_i^2 = \{m_1^i, \ldots, m_q^i\}$$

A random cross point $s$ ($1 \leq s \leq w, 1 \leq s \leq q$) is selected, determining the portion of seed nodes to be interchanged between both chromosomes. The combined chromosomes (denoted with an asterisk) are as follows:

$$c_i^{1\,*} = \{n_1^i, \ldots, n_s^i, m_{s+1}^i, \ldots, m_q^i\}$$
$$c_i^{2\,*} = \{m_1^i, \ldots, m_s^i, n_{s+1}^i, \ldots, n_w^i\}$$

Thus, the resultant combined individuals are[4]:

$$ind_1^* = \left\{ \left\{ n_1^1, \ldots, n_a^1 \right\}, \ldots \left\{ n_1^i, \ldots, n_s^i, m_{s+1}^i, \ldots, m_q^i \right\}, \ldots, \left\{ n_1^k, \ldots, n_c^k \right\} \right\}$$

$$ind_2^* = \left\{ \left\{ m_1^1, \ldots, m_d^1 \right\}, \ldots \left\{ m_1^i, \ldots, m_s^i, n_{s+1}^i, \ldots, n_w^i \right\}, \ldots, \left\{ m_1^l, \ldots, m_f^l \right\} \right\}$$

$$(5.2)$$

It is important to remark that our recombination process does not finish after simply interchanging seed nodes among chromosomes. As commented, seeds of chromosomes that cluster a hierarchical level higher than the ground one represent clusters previously created, and also they can participate in the clusterization of higher levels. That is, a seed node of a chromosome represents a hierarchy which should be interchanged entirely when combining individuals. The following example illustrates the above points.

In the individual $i=\{c_1,c_2,c_3\}=\{(2,8,10),(1,3),(2)\}$ already represented in Fig. 5.6, the seed (3) of $c_2$ refers to both the cluster generated by the third seed of $c_1$ (10) and, at the same time, it is referred by the single seed of chromosome $c_3$ (2). Thus, for instance, moving the seed node (10) of $c_1$ during

---

[4] Notice that added seeds may refer to nonexistent clusters within the receptor. In that case, those invalid seeds are not considered in the recombination.

recombination should require moving the whole structure generated by (10) to the receptor. For example, using the ground level depicted in Fig. 5.6, the result of combining the individual $i$ with another one $h=\{c_1,c_2\}=\{(3,13),(1)\}$ (see Fig. 5.8b), where their first chromosomes are combined and the cross point is set to 1, is performed as follows.

Let the original individuals be:

$$h = \{c_1^h, c_2^h\} = \{(3, 13)\,(1)\}$$
$$i = \{c_1^i, c_2^i, c_3^i\} = \{(2, 8, 10), (1, 3), (2)\} \tag{5.3}$$

By equation (5.2), the combination of the selected chromosomes $c_1$ from both individuals is:

$$c_1^{h^*} = (3, 8, 10)$$
$$c_1^{i\,*} = (2, 13)$$

and the resultant crossed individuals (shown in Fig. 5.8) are:

$$h^* = \{c_1^{h^*}, c_2^{h^*}, c_3^{h^*}\} = \{(3, 8, 10), (1, 3), (2)\}$$
$$i^* = \{c_1^{i\,*}, c_2^{i\,*}\} = \{(2, 13), (1)\}$$

Notice that after the recombination process, new seeds and an extra chromosome have been added to the individual $h$, while $i$ has lost one. This is because the interchanged seed (10) from the individual $i$ involves upper hierarchical levels, since it is referred by chromosomes $c_2^i, c_3^i$. Following the pseudo code of Fig. 5.7, function *ContainReferenceToSeed* indicates whether the clusterization of a hierarchical level uses a cluster from lower levels as a seed. Thus, in this example, both functions *ContainReferenceToSeed*$(c_2^i, c_1^i, 10)$ and *ContainReferenceToSeed*$(c_3^i, c_1^i, 10)$ return true since $c_2^i$ and $c_3^i$ both refer to the seed (10)[5] of $c_1^i$.

After recombination, the individual $h^*$ should inherit the hierarchical structure supported by the seed (10) encoded in the original individual $i$. Therefore, their chromosomes should contain proper references to the new seeds added (function *AddReferenceToSeed* in the pseudo code of Fig. 5.7), and also references to the lost seeds that should be removed (through function *RemoveReferenceToSeed*), as it occurs in the individual $i$.

### 5.4.4 Individual Mutation

Mutation is a 1-ary operation that imitates a sporadic change in individuals. Although recombination is important to maintain the best characteristics

---

[5] This reference can be direct as it occurs in the case of *ContainReferenceToSeed*$(c_2^i, c_1^i, 10)$, or indirect as in *ContainReferenceToSeed*$(c_3^i, c_1^i, 10)$, in which $c_3^i$ contains the seed (2) that refers to the 2-nd seed of $c_3^i$ (3) which finally refers to the 3-rd seed of $c_1^i$ (10).

```
PROCEDURE Recombination (Individual i, Individual h):Individuals
c=SelectChromosomeIndex(i,h)
p=SelectCrossPoint(i(c),h(c))
Seed1=SeedsToBeExchanged(i(c),p)
Seed2=SeedsToBeExchanged(h(c),p)
i(c)=i(c)-Seed1
h(c)=h(c)-Seed2
i(c)=i(c)+Seed2
h(c)=h(c)+Seed1
FOR j= 1 to size(Seed1)
                FOR k=1 to NumberOfChromosomes(i)
                   IF(c!=k && ContainReferenceToSeed(i(k),i(c),Seed1(j)))
                        AddReferenceToSeed(h(k),h(c),Seed1(j))
                        RemoveReferenceToSeed(i(k),i(c),Seed1(j))
                   END
                END
END
FOR j= 1 to size(Seed2)
                FOR k=1 to NumberOfChromosomes(h)
                   IF(c!=k && ContainReferenceToSeed(h(k),h(c),Seed2(j)))
                        AddReferenceToSeed(i(k),i(c),Seed2(j))
                        RemoveReferenceToSeed(h(k),h(c),Seed2(j))
                   END
                END
END
RETURN i,h
```

**Fig. 5.7.** Pseudocode for recombination of individuals. After the selection of the chromosomes and the seeds to be exchanged, references to those seeds within the input individuals are detected in order to be updated (in the two loops "FOR"). The *ContainReferenceToSeed* function checks whether a given chromosome contains a reference to a seed from another chromosome, while *AddReferenceToSeed* and *RemoveReferenceToSeed* are used to insert/erase seeds that refer to clusters previously generated by seeds of other chromosomes

of individuals over generations, mutation is necessary in order to avoid the stagnation of solutions in a local minimum.

In our approach we have followed a simple criterion to perform individual mutation, which consists of eliminating/adding randomly a seed from/to a random chromosome of the individual to be mutated. More formally, the mutation of a particular chromosome $c$ that contains $s$ seeds, defined as $c = (n_1, \ldots, n_s)$, may result in one of the following possibilities:

$$\begin{aligned}
c_{add} &= (n_1, \ldots, n_s, n_{s+1}) \\
c_{del} &= (n_1, \ldots, n_p, n_{p+2}, \ldots, n_s)
\end{aligned} \tag{5.4}$$

As notices when discussing the recombination process, the effect of eliminating/adding seeds may also cause relevant changes in the structure encoded by

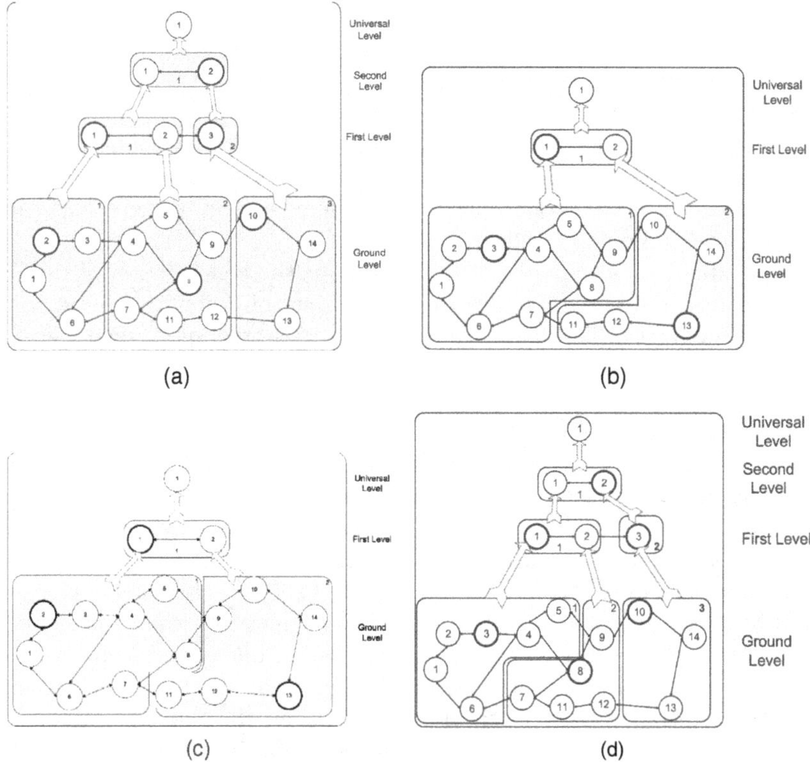

**Fig. 5.8.** Example of individual recombination: (a)(b) Hierarchies encoded by the individuals $i$ and $h$ referred in the text; (c)(d) Resultant offspring $i^*$ and $h^*$ after crossing as commented in the text. Notice how recombination combines individual characteristics such as the number of hierarchical levels

individuals. For instance, the deletion of a particular seed from a chromosome may modify the number of levels of the hierarchy represented by the individual. This effect can be observed in the following example, where the individual $i$ undergoes the deletion of the first seed of the chromosome $c_2$, which leads to the elimination of $c_3$:

$$i = \{c_1, c_2, c_3\} = \{(2, 8, 10), (1, 3), (1)\}$$
$$i^* = \{c_1, c_2\} = \{(2, 8, 10), (3)\}$$

(5.5)

Regarding the effect of inserting a random seed into a chromosome, it will only increase the number of clusters at certain levels, but no changes will be produced in the number of levels of the hierarchy represented by the individual.

Notice that the random mutation of individuals (as well as the recombination) may not yield good results in the sense that the offspring is not well adapted to the problem at hand. We rely on the evolutionary process that will eliminate weak individuals through their fitness.

### 5.4.5 Individual Evaluation

Evolution of the population (AH-graphs) is carried out through the typical genetic operations, recombination and mutation, previously discussed, after a selection process. This selection process, which takes into account the *fitness value* of each individual (that is, the goodness of each AH-graph for improving the situatedness of the robot), has been implemented by using the so-called *roulette wheel* technique[6] [158].

The evaluation of individual fitness involves the construction of the AH-graphs represented by individuals to test their suitability with respect to the current robot tasks. The hierarchy constructor submodule (recall Fig. 5.1) uses the information encoded by chromosomes to construct hierarchies based on the clustering algorithm of Fig. 5.5. Each constructed AH-graph is then evaluated considering the effort of planning the robot tasks as well as an heuristic estimation of their execution. For assessing the fitness of individuals we rely on an efficient task-planning approach called *HPWA* (Hierarchical Planning through World Abstraction) which has been previously presented in [56]. Broadly speaking, this approach takes advantage of the hierarchical structure of the world model by solving the task at a high level of detail, and then uses the resulting abstract plan to ignore irrelevant elements at lower levels with respect to the current goals. In our experiments we have chosen Metric-FF [76] as the embedded planner within the HPWA scheme. Apart from optimizing the internal representation, the Hierarchy Optimizer also updates the estimate of the probability distribution (frequency of occurrence) of the tasks over time[7]. With all that information, the fitness of a given hierarchy $H$ is computed by:

$$Fitness(H) = \frac{1}{\sum_{i=0}^{n-1} freq(t_i) \cdot (p(t_i) + h(t_i))} \tag{5.6}$$

This function gives the fitness (goodness) of individuals for planning and executing the current set of robot tasks. In this expression, $n$ is the number of tasks requested to the robot up to the current moment, $freq(t_i)$ is the frequency of occurrence of the task $t_i$, $p(t_i)$ is the planning effort to solve task $t_i$ (planning time) and $h(t_i)$ is an estimate of the cost of executing the task. This execution cost estimate is based on the experiences of the robot in past executions of the task. In our case, we average time consumption, although other estimates can be used, e.g., energy consumption, mechanical stress, or a mixing of several factors, as it is done for route planning in [45]. Notice

---

[6] Efficiency has not been our concern here, but simplicity. Instead of this *relative fitness* selection, other types of selection techniques can be used, such as *rank-based* or *tournament* methods [9].

[7] Other frequency estimates could be developed, for example, based on Bayes filtering [1].

that including an estimate based on past executions in $Fitness(H)$ avoids the physical performance of tasks for each individual of the population, which would be unacceptable for on-line optimization.

By considering the cost of planning and execution, the current set of robot tasks, and their frequency of occurrence, the optimization process guides the search to find those symbolic models of the environment better adapted to solve frequent tasks in detriment of other infrequent ones. Notice that the final constructed model can be far from the optimal at first, but our approach guarantees its improvement over time, as it is demonstrated in the experiments of Sect. 5.5.

### 5.4.6 Memory and Time Consumption

In this section, some details regarding memory consumption and computational complexity of our algorithms are given.[8]

#### Memory Consumption

The execution of the genetic algorithm only keeps in memory a set (population) of chromosomes at a given iteration (set in our experiments to *15*, which proved to be sufficient for convergence) and the set of the tasks to be performed. On the one hand, considering the encoding of chromosomes, that is, sequences of seeds (integer numbers), for hierarchies with a ground level with $n$ nodes, and hierarchical levels that contain half the nodes of the next lower levels, the number of hierarchical levels becomes $log_2 n$ and the number of seeds needed to represent each of these hierarchies is $\sum_{i=1}^{log_2 n} \left(\frac{n}{2^i}\right)$, that is, $O(n)$. In the worst case, in which each hierarchical level reduces only one node, the number of levels is $n-1$ and the number of seeds needed to represent each hierarchy is then $\sum_{i=1}^{n-1}(n-i) = \frac{n(n-1)}{2}$, that is, $O(n^2)$. On the other hand, the information kept in memory for the set of tasks to be performed is negligible: it consists of the definition of the goal to be achieved, i.e. *(at box-1 room-9)*, the frequency of occurrence of each task (a float number), and an averaged value of the execution time of the task (a float number).

#### Computational Complexity

Within the genetic algorithm, chromosomes are constructed at every generation to calculate their fitness. For the clustering algorithm (see Fig. 5.5), let $r$ be the number of seeds of a given chromosome that cluster $n$ nodes from the lower hierarchical level, and let $c$ be the average connectivity of a set of nodes. In the worst case, in which $r = n-1$ and $c = n-1$, the complexity of the algorithm is $O(n^2)$ due to the two inner loops. The outer loop (the 'while' loop)

---

[8] The complexity study of the anchoring and planning processes is left out of the scope of this book.

ensures that all nodes are clustered, so we can bound the clustering process in the worst case as $O(n^3)$. As this process is repeated per each chromosome of every individual of the population, the computational complexity of calculating the fitness of a population with $m$ individuals becomes, in the worst case, $O(m(n-1)n^3)$, which is $O(n^4)$ when $m$ is a relatively small constant with respect to $n$.

The recombination process (see Fig. 5.7) consists of exchanging seeds between two chromosomes and then updating their references in the upper ones. The complexity of this process is clearly polynomial and it can be bounded as follows. Let $s_1$ and $s_2$ be the number of seeds to be exchanged. The first part of the algorithm (select, insert, and remove elements from a list) is polynomial with respect to the number of seeds to be exchanged, that is, $O(max(s_1, s_2))$. The updating of seeds consists of two loops based on $s_1$, $s_2$, and on the number of the upper chromosomes of the individuals, which is $n$ in the worst case, as explained in the previous section. Thus, the computational complexity for each part is $O(s_1 n)$, that is $O(n^2)$ in the worst case, being the called functions (*Contain-*, *Add-*, and *Remove-ReferenceToSeed*) linear on the size of the seeds of the chromosome. Thus, the recombination process is $O(\frac{m}{2}n^2)$, considering that each individual is combined with another only once.

Finally, for the study of the computational complexity of the mutation process (see Fig. 5.9) only the update of upper chromosomes is considered, and thus, it can be approximated in the worst case as $O(n)$.

```
PROCEDURE Mutation (Individual i):Individual
type=RandomSelectTypeofMutation()
c=ChromosometobeMutated(i)
IF (type==Add)
                sn=RandomSeedNode()
                i(c)=i(c)+sn
ELSE

                sn=RandomSeedNode(i(c))
                i(c)=i(c)-sn
                FOR k=c+1 to NumberOfChromosomes(i)
                     IF (ContainReferenceToSeed(i(k),i(c),sn)
                          RemoveReferenceToSeed(i(k),i(c),sn)
                          IF (IsEmpty(i(k))) RemoveChromosome(i(k))
                     END
                END
END
RETURN i
```

**Fig. 5.9.** Pseudocode for individual mutation. When a seed is deleted, all references to it are deleted too. When adding/removing seed nodes from a certain chromosome, references from higher ones must be reordered to correctly keep previous references (the two inner 'for' loops)

## 5.5 Experimental Results

We have evaluated the performance the ELVIRA framework when planning/executing robot tasks. For that, we have conducted a variety of simulations and real experiments, both aimed to test the hierarchy optimization process within a real scenario (Sect. 5.5.1), and to test the behavior of our software system in larger and specially complex environments (Sect. 5.5.2). Section C gives a brief discussion about the applicability of our approach to other types of scenarios (outdoor or dynamic).

### 5.5.1 Real Evaluation

For real experiments, ELVIRA has been integrated into a robotic system that runs an implementation of the *Thinking Cap* hybrid architecture, suitable for controlling situated agents [129]. Our experiments have been carried out on a Magellan Pro robot equipped with a ring of 16 ultrasonic sensors and a color camera which provide sensor information to the Anchoring Agent in order to anchor distinctive places for robot navigation and objects (colored boxes of a given size).

The test scenario is the office environment depicted in Fig. 5.10, made up of several rooms and corridors. It should be noted that although this is not a representative large-scale environment, it is large in fact, since we are considering planning tasks involving not only navigation but also object manipulation, which may easily become an intractable problem [56] (large-scale scenarios are considered in the experiments in Section B).

In this experiment, the robot performed within the test scenario during 8 hours of a non-working day (in three consecutive runs of 2 hours separated by 1 hour for battery recharge). The experiment starts with an initial ground level consisting of a graph of 128 distinctive places (20 rooms/areas), their connections, and 5 boxes placed at certain rooms (see Fig. 5.10). Within this initial setup the robot is requested to roam while anchoring continuously new distinctive places and colored boxes found in the environment. In our particular implementation, the Hierarchy Optimizer works to improve the current model with a cycle period of 1 min, which is an upper bound that in our implementation ensures the termination of each optimization step. In parallel, the robot is commanded every 5 min to execute a random task (out of a set of tasks also chosen at random) using its current internal representation of the world. At the end of the experiment, the robot has incorporated into the symbolic model 256 new distinctive places (corresponding to 33 rooms), more than 750 connections between them and 40 boxes.

Initially we have considered a set of 3 random tasks that involve world elements from the initial ground level given to the robot. Once the robot has explored the major part of the environment (after approximately 40 min), 3 new random tasks are added to the set of robot tasks that involve the new anchored information. The type of tasks considered consists of moving a

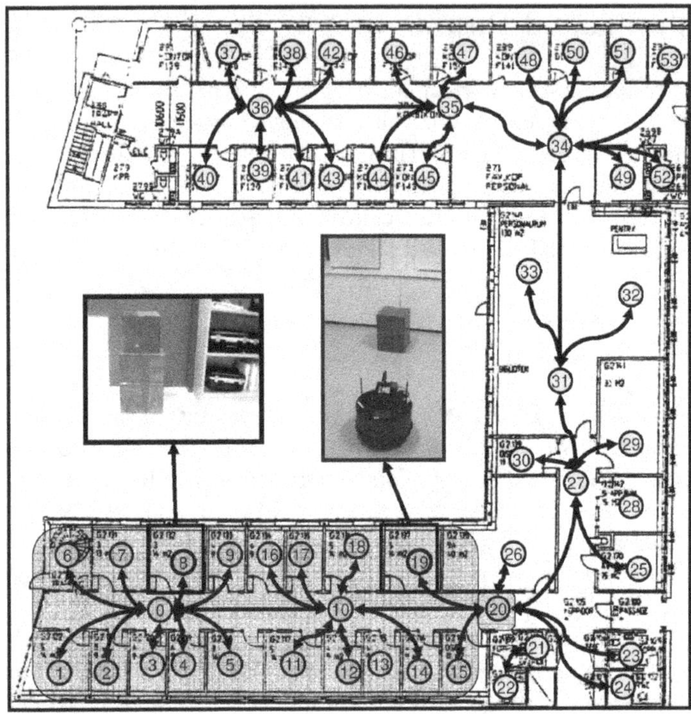

**Fig. 5.10.** Real scenario for our robotic experiments: a part of the Mobile Robotic Lab at the AASS Center in the University of Örebro (Sweden). The shaded region represents the initial set of rooms given to the robot, while the rest has been anchored in successive explorations (for clarity, nodes that represent objects and distinctive places inside rooms are not depicted). In the middle of the figure, we show the robot used in this experiment and the appearance of a typical room

certain box from one location to another, possibly requiring the manipulation of other boxes, e.g., to unstack boxes which are on top of the desired one. More precisely, they include a simple task (Task 3 in Fig. 5.11) that only consists of navigating from room 6 to room 14 (see Fig. 5.10), and another five for which the robot has to make a plan for manipulating (unstack/stack/pick up) a box from a heap and moving it to another room to placed it on a pile.

Since the chosen robotic platform can not manipulate objects, the robot physically executes only the navigation between the locations involved in the tasks, being the manipulation carried out by a person (although the task-planning process deals with both operations automatically[9].) Both the

---

[9] Since our concern here is the optimization of the internal representation of the robot, possible navigational failures as well as errors in anchoring are not considered in the results presented in this section. In both cases, the robot failure was manually recovered by an operator.

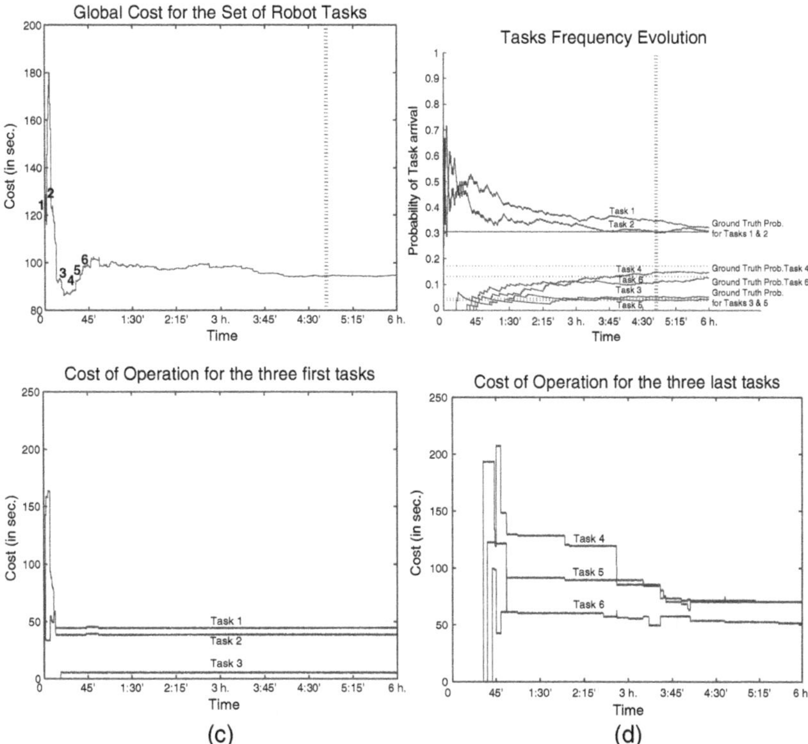



**Fig. 5.11.** Model optimization in a real experiment. a) Evolution of the global cost for the set of robot tasks, measured according to equation (5.6). Arrows on the plot indicate the first time that each task is commanded to the robot. Notice how the cost oscillates substantially while the robot has not enough knowledge about its tasks nor the environment. As long as the robot optimizes the symbolic model, oscillation diminishes until a steady-state is reached around the 5th hour of operation. b) Frequency of occurrence of each task perceived by the robot and the ground truth probability of occurrence of tasks (horizontal lines). c-d)Evolution of the cost of solving and executing each robot task along the whole experiment. Notice the overall improvement in planning/execution

time required to perform planning and topological navigation are considered to guide the model optimization through the function cost (5.6).

Figure 5.11a measures the optimization achieved in the model during the experiment. Observe that after the high raise in the cost of operation at the beginning, which corresponds to the request of a manipulation task (in room 8) and the inclusion in the model of new anchored information (a pile found in that room with three new boxes), the optimizer achieves a decrease on the global cost. Thus, after approximately 40 min, the internal symbolic model is adapted to the initial set of tasks and to the environment. After that point, raises in the figures are caused by the inclusion of 3 new manipulation tasks

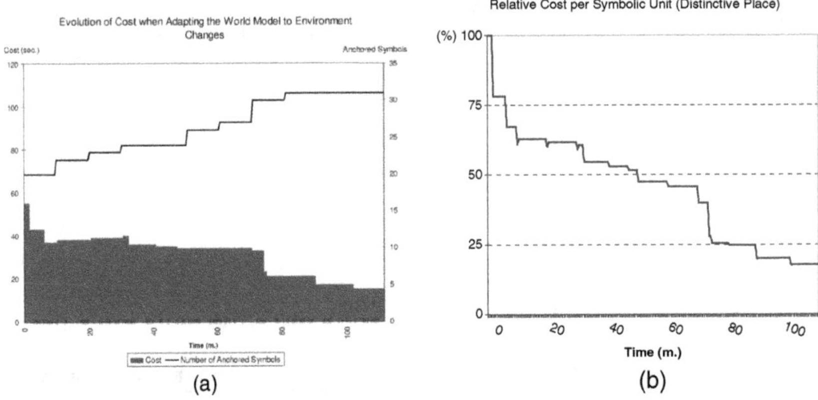



**Fig. 5.12.** Optimization of the AH-graph model under world changes. These charts show the evolution of the cost of operating over time when genetic individuals (AH-graphs) are adapted to changes. a) Observe how the cost only increases slightly when changes are perceived. This is because the robot does not lose completely the experience acquired up to the moment when these changes are detected. Chart b) plots the same results that chart a) but with respect to the number of anchored symbols, normalized by the initial cost, that is, $f(t) = \frac{cost(t)}{places(t)} \cdot \frac{100}{\frac{cost(0)}{places(0)}}$

and also by the inclusion in the model of new anchored information. As long as the robot does not perceive changes in the environment, that is, no new anchored information is added to the internal model of the environment (after the 5-th hour), and the estimated frequency of the requested tasks approximates to their actual probability of occurrence (as shown in Fig. 5.11b), the overall cost of the robot tasks reaches a steady state. This occurs after around 8 hours from the beginning of the experiment, approximately.

In our experiences we have also demonstrated that the Hierarchy Optimizer approach preserves the knowledge acquired by the robot along its operational life. That is, it does not construct a new symbolic model from scratch each time world changes. For illustrating that, we consider an experiment similar to the one described before, but the robot is now commanded to solve only one task within a reduced part of our test scenario. Results of this experiment are shown in Fig. 5.12. Figure 5.12a displays the evolution of the cost of operating within the environment, in which the number of distinctive places is progressively incremented. Notice how the increase in the world complexity (the number of distinctive places passes from 20 to more than 30 at the end of the experiment) does not affect the performance of the Hierarchy Optimizer. Figure 5.12b refers to the same results but measuring the cost with respect to the number of places anchored at each moment. This chart shows how the average operation cost per place has been reduced around 75% over time, despite the growth in the complexity of the symbolic representation.

### 5.5.2 Simulated Experiments

Due to the practical limitations involved in real scenarios, we have also conducted several simulated experiments in order to test our approach in larger environments. The design of our simulations are based on results previously presented in [56] and commented in Chap. 3. There we state the low performance of general planners like Metric-FF even when dealing with relatively small scenarios for which the spent time for planning grows rapidly, even making the problem intractable (recall Fig. 3.11).

Our simulations here aim to prove two issues: a) that the quality of the resulting plans obtained by ELVIRA is similar to the ones obtained by general planners, and b) that ELVIRA can deal with and adapt its behavior to changing environments, which can not be faced by general planners.

For a), we have chosen a normal-size environment (256 world elements) and compared the results obtained by ELVIRA and Metric-FF (see Sect. 5.5.2). For b), we have chosen a larger environment with more than 1024 symbols (96 rooms) to test the adaptation yielded by ELVIRA along time (see Sect. 5.5.2). Notice that although these scenarios can not be considered as spatially large-scale, they are complex enough to prevent general planners to deal with them.

**Simulated Experiment 1**

The first of our simulated experiments is meant to prove the quality of the plans found by ELVIRA during the optimization process. For doing so, we have simulated an environment with a maximum of 256 elements (places + objects) where five different tasks are considered for planning. Figure 5.13a shows the evolution of the average plan length (in number of actions) for both the Hierarchical Planner using the optimized symbolic model and flat planning using the information at the ground level. Note that, in both cases, the average plan length is quite similar and that the small discrepacy between then is acceptable when considering the difference in the planning costs depicted in Fig. 5.13b.

Figure 5.13b shows the task-planning cost for this experiment. The hierarchical planning in the optimized hierarchy yields remarkably better results than planning only with the ground level information. Note that, for both cases, variations over time are due to the arrival of new symbolic information (anchoring) and to the effect of the task frequency in the computation of the cost function.

**Simulated Experiment 2**

In a second type of experiments we have simulated larger environments (up to 1024 ground symbols) which can not be solved by general planners. With these experiments we prove the suitability of ELVIRA for providing an efficient task-planning process within a changing and relatively large environment.

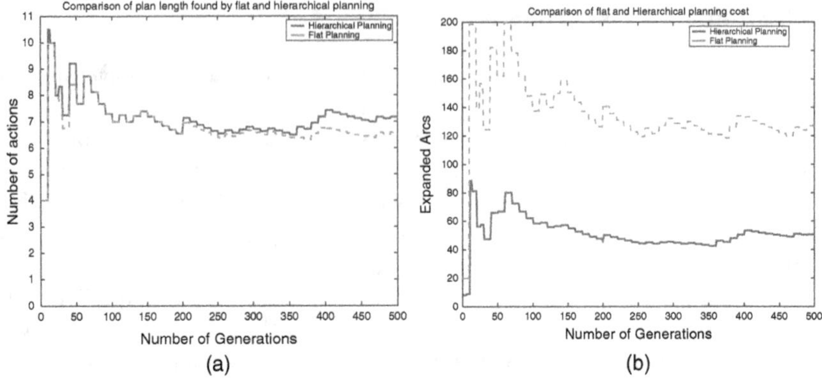

**Fig. 5.13.** Results of a simulation in a medium-size scenario: (a) Evolution of plans quality (number of actions) along the simulation. Notice that although plans yielded by the hierarchical planning are not always the optimum, in average there is a slight difference with respect to the quality of plans found by the non-hierarchical planner; (b) Evolution of planning cost. Notice the good performance of hierarchical planning with respect to the cost of planning with non-hierarchical information. Also note that ELVIRA adapts the hierarchical model according to changes in the environment (anchoring and new tasks), which is shown on the smooth evolution of the hierarchical planning cost

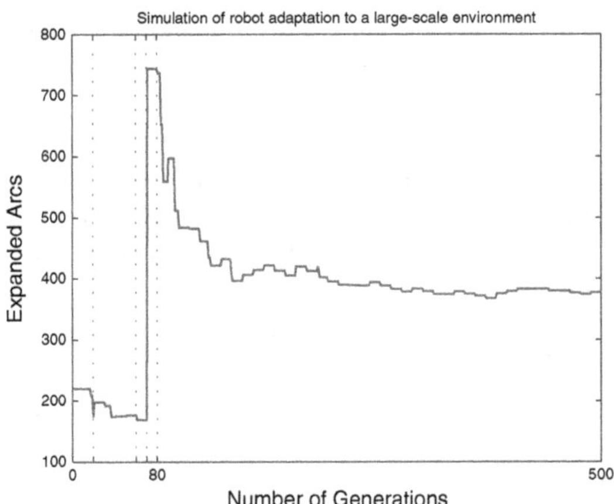

**Fig. 5.14.** Planning cost evolution within a large environment. The arrival of a new task not previously considered by ELVIRA is marked with a dotted line. Note the high peak at generation 70, which is due to the request of planning a complex task. After that, observe how ELVIRA adapts the hierarchical model with respect to the five considered tasks, which are continuously requested each 10 cycles

Figure 5.14 shows the evolution of the planning cost for an experience with five random tasks (planned every ten cycles) within a variable environment with up to 1024 ground symbols (places and objects). Note that after cycle

80 in which ELVIRA has planned all the considered tasks, the Hierarchy Optimizer tends to decrease the planning cost significantly, until a steady-state situation is achieved at the end of the simulation.

### 5.5.3 Considerations for Outdoor or Dynamic Environments

The approach presented above is mainly intended for, and it has been tested in, indoor environments that have a moderate amount of dynamism. However, we believe that the core ideas of our long-life optimization mechanism could extend beyond these limits, notably to outdoor environments and/or to more dynamic environments.

The application of our approach to outdoor scenarios is possible in principle, provided that a symbolic model of the outdoor environment of interest can be defined. This implies that the environment has enough structure to be amenable to a symbolic representation. However, some thorny problems related to perception should be tackled in this case. Indoor scenarios exhibit some characteristics that provide the symbols that represent the environment with more stability. For instance, light conditions that are of vital importance for visual anchoring of objects are more easily controlled in indoor scenarios. Outdoor applications may require more sophisticated anchoring techniques, like the one presented in [36]. Moreover, indoor environments like the one used in our experiments are usually structured in rooms, corridors, etc, which permits the robot to construct maps and extract the type of spatial features (rooms and distinctive places) used in our experiments. For outdoor applications, other features and localization techniques should be considered, as well as different sensors such as infrared cameras or GPS, albeit the symbolic part of our approach should not suffer major changes.

For what concerns highly dynamic scenarios, it should be noted that our approach currently assumes a moderately dynamic environment: objects may be added or removed, but they are otherwise expected to be static. A more dynamic environment would pose two types of problems to the current approach. First, we would need to rely on a more sophisticated anchoring process, able, for instance, to track the position of moving objects, in order to avoid errors in the creation and update of anchors. These errors would produce a great variability in the symbolic model, which might cause the symbolic optimization process lose its convergence. A more sophisticated anchoring process, with advanced tracking capabilities, would also help to cope with cluttered environments. Second, our evolutionary algorithm requires a number of iteration to achieve an acceptable solution. Each time world changes (the ground level of AH-graphs), the algorithm has to face variations on the problem to be solved, and a new period of time is required to attain a solution. Currently, we assume that after a period of time, in which the scenario may gradually change, the environment remain static and unalterable. Further research needs to be conducted to cope with more dynamic situations. However, neither the hierarchical representation of the space nor its optimization process should undergo fundamental changes.

## 5.6 Conclusions

In this chapter we have described a framework to cope with a problem of great importance in mobile robotics: the construction of near-optimal symbolic models of large environments for a mobile robot. Our system has been implemented as a software we call ELVIRA, which has been integrated into a real robot. Our approach considers three issues: (i) the maintenance of the symbolic model anchored and updated with respect to the environmental information, (ii) the structuring of the model for efficient processing in the presence of large amount of information, and (iii) the optimization of such model with respect to the tasks at hand, to improve over time the robot performance.

We rely on abstraction through the mathematical model AH-graph in order to cope with large environments. The creation of the ground level of the AH-graph is achieved automatically by using an anchoring algorithm that detects rooms or distinctive places for robot navigation, and objects. Upon that ground symbolic information, an evolutionary algorithm is responsible for creating a near-optimal hierarchical structure with respect to environmental changes and variations in the robot operational needs. Although evolutionary algorithms are not usually considered in dynamic scenarios, our experiments have shown that when their work is spread over time (considering them as any-time algorithms) they exhibit good results in optimization.

# 6

# Implementation and Experiences on a Real Robot

*What we plan we build.*
Conte Vittorio Alfieri, dramatist (1749–1803)

Once we have devised efficient mechanisms for robot operation and human communication through multiple abstraction, it seems we are now on the brink of developing a robotic system like our robotic taxi driver. Unfortunately there is still, at least, a hard obstacle to be solved: robots can not overcome all possible situations in a real scenario. That is, robot skills required to perform in a complete autonomously manner within a dynamic, unpredictable, human scenario are still beyond the capabilities that the current robotic technology can offer. This fact is what actually prevents the presence of robots around us.

How can we enable robots to overcome this situation? Restricting our view to robots that closely work with humans, we could approach the problem by devising a robotic system capable to require human help in order to accomplish certain goals. Human participation in the robot mission can make up for the current lack of robot autonomy: for instance, a sweeper robot intended to clean a certain room could require human help to open the entrance door.

Following this idea, this chapter describes a robotic architecture, called *ACHRIN*, specifically designed to enable a mobile robot to closely work with people. It implements all tools described in previous chapters, and furthermore, it considers mechanisms to detect robot inabilities or malfunctions when performing some operation in order to ask for human help.

ACHRIN has been initially designed for assistant robotic applications like a robotic wheelchair, due to the close tie between human and machine in that case; however, it could be extended to other types of applications. In fact, you can consider such a vehicle (a robotic wheelchair) plus its control system (ACHRIN) as an instance of our taxi and its futuristic robotic driver. This chapter ends with a description of the robotic wheelchair SENA and some real experiences carried out on it.


C. Galindo et al.: *A Multi-Hierarchical Symbolic Model of the Environment for Improving Mobile Robot Operation*, Studies in Computational Intelligence (SCI) **68**, 115–141 (2007)
`www.springerlink.com`                                       © Springer-Verlag Berlin Heidelberg 2007

## 6.1 Introduction

Robotic presence in our daily life is still limited. Unfortunately, there are neither servant robots helping us in hard tasks, nor robotic taxi drivers taking us home. One of the most important reasons of this is that full autonomous performance of mobile robots within real scenarios is not possible yet, mainly due to difficulties in coping with highly dynamic environments, treating with uncertainty, sensor and/or software robustness, limited robotic abilities, etc.

Along this book we have devised mechanisms to improve robot reasoning and robot-human communication. Now we present a direct and practical approach to reduce the evident lack of robot autonomy in certain (many) applications: considering that the robot is supposed to work closely to people (the so-called *assistant robots*), we will enable machines to ask for human help. Thus, we permit humans to complement the robot's skills needed to accomplish its tasks. People present in the robot environment can *physically* help the machine by both extending robot abilities (for example to open a locked door) and/or improving them (for example, the human can perform a flexible and robust navigation by manually guiding the vehicle). This scheme makes necessary the human participation at each level of the robotic system, ranging from a *high level*, i.e. advising a route to arrive a destination, to a *low-level*, reporting information as an extra sensor (i.e. the human can work detecting close obstacles).

Although in our scheme people help the robot to overcome certain situations, the assistance offered by robots is normally relevant enough to assume this human participation into the robotic system (especially in the case of assistant robots). Thus, for example a robotic fair host can guide people through different stands showing information during long periods of time, though it could need human help to self-localize if it occasionally fails. Another clear example is a *robotic wheelchair* (as the one described at the end of the chapter (Sect. 6.7.1)) which provides mobility to elderly or impaired people relieving her/hir of approaching tedious tasks like moving through a crowded corridor. In that example, possible robot limitations, such as opening a closed door, are negligible considering the benefits that the user obtains from the machine. Thus, we can consider that both, human and robot, improve the autonomy of each other reaching to a unique system (human+robot) which takes advantage of their separate abilities.

What we propose in this chapter is therefore a *human-robot integration* that improves the autonomy of the whole human+robot by human participation at all levels of the robot operation, from deliberating a plan to executing and controlling it.

In the robotic literature, human participation in robot operation is not a new idea. Some terms have been coined to reflect this, such as *cooperation* [63, 143], *collaboration* [54], or *supervision* [54, 131]. In all these cases the human is an external actor with respect to the robotic system, who can only order robot tasks or supervise its work. We take a further step by *integrating* the human into the system, considering her/him as a constituent part of it.

To approach the proposed human-robot integration, we identify and address the following goals:

(a) *To manage knowledge about the human physical abilities.* The system must be aware of the physical abilities of the human in order to decide how and when they can be used. For example, in a given robotic application, people can manipulate objects (open a door or call a lift) and perform manually navigation through some mechanism (like a joystick), while the vehicle can perform navigation, maybe through different navigational algorithms. Thus, human integration may provide new abilities, and therefore, the robotic system must be able to consider them when planning/executing a task. In our current scheme we implement a simple selection process to choose abilities from the available human-robot repertory that consists of selecting human abilities only when the robot is not able to perform them.

(b) *To consider the human perceptual abilities.* Robot sensors can not capture reliably either the highly dynamic nature of real scenarios or some risky situations (for instance, glass doors are not detected by some robot sensors, like lasers). The integration of a human into the robotic system may extend the robot autonomy by permitting her/him to work as an intelligent sensor that reports or predicts dangerous situations.

(c) *Detection of execution failures.* The robotic system must detect whether the execution of the current action has failed. The recovery action may include the human: for example a navigational error can be detected, inquiring human help to guide the vehicle to a safety location. For that purpose, the human could improve the robot's capacity of detecting those failures.

(d) *High-level communication.* Finally, active human integration within the robotic system obviously requires a high-level communication mechanism enabling the user and the robot to interact in a human-like manner, i.e., using natural language, with human concepts which involves symbols such as "door", "room", "corridor", etc.

For addressing these goals, we have designed and implemented a specific control architecture, called *ACHRIN -Architecture for Cognitive Human-Robot Integration-* (Fig. 6.1) that copes with the above mentioned issues: (a) it enables the task planning process (HPWA in our case) to consider human actions, and permits the user to sequence and execute them at the lower levels of the architecture; (b) it permits the user to verbally report collision alerts at any time; (c) it both checks regularly the performance of the vehicle and enables the user to report failures. Finally goal (d) is approached by the use of commercial speech recognition and voice generation software [150], [154], as well as a cognitive hierarchy within the multi-hierarchical world model of the robot that serves as a suitable interface with the human cognitive map (as explained in Chap. 4).

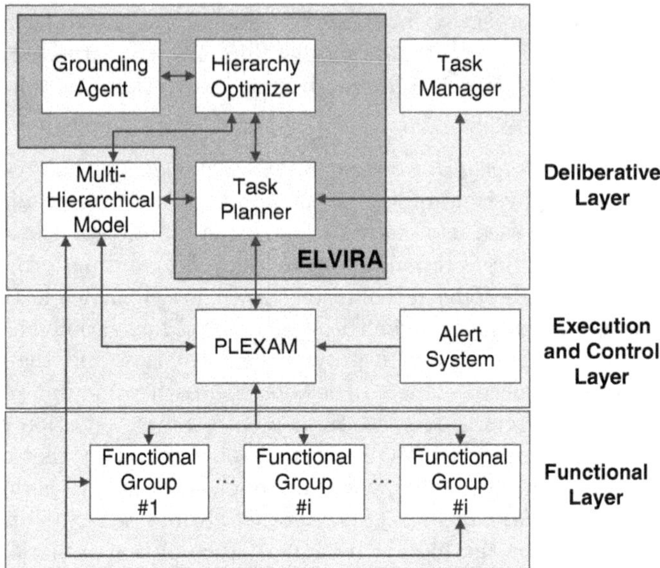


**Fig. 6.1.** A general view of ACHRIN. Broadly, it can be considered as a hybrid robotic architecture. However, it does not fit strictly into that typical hierarchical arrangement. For example, the World Model module can be accessed by modules of all the layers

Human-robot integration is materialized by designing each component of ACHRIN through a common structure called *CMS* (*common module structure*), which is able to encapsulate humans as software algorithms that provide certain results after being executed. Each CMS groups and manages *skills* (carried out either by the human or the robot) which are aimed to achieve a similar goal. For instance, one of the CMS of ACHRIN is the *Navigational CMS*, that entails different skills to move the vehicle between two locations: a variety of robotic skills, like reactive navigation or tracking a computed or recorded path, and a human skill consisting of manually guiding. These skills are invoked and managed by ACHRIN in the same manner without distinction of the agent that performs the action, human or robot. The human integration we have achieved with ACHRIN enables the user to perform low-level navigation tasks like manoeuvering in complex situations, as well as high-level tasks, like modelling symbolically the workspace, as shown further on.

In the following, Sect. 6.2 reviews some previous works related to the interaction between humans and robots. An overview of ACHRIN is given in Sect. 6.3, while its components are presented in detail in subsequent sections. A particular instantiation of ACHRIN for a real mobile robot, in particular to a robotic wheelchair, is described in Sect. 6.7.

## 6.2 Related Works on Human-Robot Interaction

Human Robot Interaction (HRI) has been largely treated in the robotics literature from different perspectives. Scholtz [131] proposes five different human roles in robotic applications (from supervisor to bystander) which cover most of the found approaches. The most common human robot interaction is to consider the human as a robot's supervisor (*supervisory control* [78, 133]). This implies that tasks are performed by the robot under the supervision of a human instead of the human performing direct manual execution.

In the teleoperation area, *collaborative control* [54, 53] can be applied, which is a particular instantiation of supervisory control. Through collaborative control, robots and human dialogue to decide the actions to be carried out by the machine. This relation between human and robot improves the robot operating capacity, but it prevents the human to physically act when the robot is not capable to continue its plan, for example when a mobile robot must pass through a closed door.

The concept of *cooperation* (see for example [63, 143]) is also well spread within the robotic community. Through human-robot cooperation, humans can perform some coordinate tasks with machines, adopting different roles ranging from coordinator, where the human (also typically an expert) only supervises the robot operation, to a role of a robot partner, where human and robot work independently to achieve a common objective. Nevertheless, these approaches only consider humans as external agents of the robotic system, but not as a constituent part of an unique human+robot system that can take full advantage of the sum of their abilities.

There is a variety of robotic architectures aimed at supporting human-robot relations of the type described before. The general trend is to consider a three-layered software architecture with the typical *deliberative, intermediate, and reactive* tiers. But what makes ACHRIN different from the rest is the level at which the human interacts with the robotic system. Most of robotic architectures implements human-robot interaction at the highest level, that is, they rely on the human planning capabilities while low-level operations, i.e. navigation, are carried out by the machine [100, 101, 95, 119]. A few works considers the human participation at the intermediate level [74, 81], or at the reactive layer [3]. But up to our knowledge, no work has approached a general framework to consider (integrate) the human participation into the robotic system at all levels of abstraction. The closest work to ours, presented in [122, 114], also focusses on the human-robot interaction for robotic wheelchairs. They also propose a human interaction at all levels of the architecture, but in a restrictive manner, without considering the wide range of abilities that the human can offer such as perception and manipulation abilities, world modeling, task-planning, etc. One of the novelties of our approach is the integration into the architecture of a cognitive spatial model through a cognitive hierarchy as the one described in Chap. 4.

The term "human integration" has been previously used in [3] in the same terms as in this chapter, but only considering human abilities at the lowest level of the architecture. That work, as all approaches classed as *shared control* systems [94, 157], combines at the same time human and robot commands to perform low-level tasks, (i.e. control the velocity of a vehicle), however, it lacks for mechanisms to allow the human to take full control of the robot, which becomes necessary in assistant applications.

Other works implement the so-called *adjustable autonomy* in which machines can dynamically vary their own autonomy, transferring decision making control to other entities (typically human users) in key situations [30, 37, 44, 130]. Since a human can take decisional tasks, ACHRIN can be seemed as an adjustable autonomy system, but in addition, it enables the human to perform physical actions.

Finally, the *Human Centered Robotics* (HCR) concept has emerged to cope with some of the specific requirements of robotic applications within human environments [37, 80, 110]. Among the different questions considered by the HCR paradigm, two important issues are *dependability* and *human-friendly interaction*. Dependability refers to physical safety for both people and the robot, as well as to operating robustness and fault tolerance. Also, HCR considers human-friendly communication, which implies the capability of easily commanding the robot as well as reporting execution information in a proper human way. ACHRIN fits into the HCR paradigm since improving mechanisms for human integration into the robotic system also strives for robot dependability and human-friendly interaction.


## 6.3 Overview of the ACRHIN Architecture

The ArChitecture for Human-Robot Integration (ACHRIN) is based on a hybrid scheme [135] made up of a number of elements[1], called *modules*, which are grouped into three layers (see Fig. 6.1):

- *The Deliberative layer* entails the ELVIRA framework for the automatic (and also manual) construction of the symbolic and multi-hierarchical that represents the robot environment, as commented in Chap. 5. It also produces plans that, within ACHRIN, may include human actions to achieve a certain goal (see Sect. 6.4).
- *The Execution and Control layer* sequences and supervises the execution of plans taking into account the information collected from the functional layer and the robot's sensors (refer to Sect. 6.5). According to such information, it may tune the behaviour of certain modules, i.e. reducing the vehicle speed when dangerous situations are detected.

---

[1] A preliminary approach of a multi-agent version of ACHRIN has been recently presented in [13].

- *The Functional layer* comprises a number of groups of skills, called *functional groups*, which physically perform actions, like navigation, manipulation, etc. (see Sect. 6.6). Each functional group may entail different ways to accomplish a particular type of action. For example, the robot can traverse between two spatial points either by a reactive algorithm, by tracking a computed path, or by the user manual guidance. In the case the robot is not capable to perform certain types of actions (as occurs in the implementation described in Sect. 6.7 for manipulation tasks), they must be carried out by a human.

To support the all-level human integration we claim here, we have used the *common module structure* (CMS) shown in Fig. 6.2 as a skeleton for all the modules in the architecture. The CMS integrates human and robot abilities through the so-called *skill units*. Each CMS may contain a variety of (human or robotic) skill units that materialize a particular type of ability, like producing a plan, checking for risky situations, moving between two locations, manipulating objects, etc.

In a deeper description, the elements that constitute the CMS are the following:

- *Skill Units.* Skill units execute the action that the module is intended to carry out. Both robotic and human skill units return to the processing core a report, indicating whether they have executed correctly or not.
- *Skill Unit Common Interface.* Although units within the same module of the architecture carry out the same action, they may exhibit differences. For example, the plan to achieve a goal may include a navigation action to a symbolic location given in terms of a planning hierarchy, like "Go to R2-2". To execute such an action, a robotic unit may need the geometric position of the spatial concept "R2-2", let say $(x = 3.4, y = 1.85, \phi = 20)$, but a human skill unit would rather need its linguistic label, i.e. "Peter's office". The Common Interface component of the CMS retrieves from the *World Model module* the information needed in each case[2], which may involve a symbol translation between hierarchies from the multi-hierarchy as explained in Chap. 4.
- *Processing Core.* It receives action requests, i.e. "navigate to a place", and invokes the corresponding skill unit to execute them. When no extra information is provided, the processing core firstly chooses a robot skill unit to accomplish the requested action (following a certain selection policy, i.e. the one with the highest level of past success). The processing core is also in charge of receiving and communicating the results of the skill unit execution to the rest of ACHRIN's modules.
- *External Communications.* This element encapsulates two mechanisms to communicate different modules of the architecture: *client/server requests*

---

[2] This is the reason of the pervasive interconnection of almost all modules of the architecture to the World Model module. The Alert module is the unique exception since its work is purely subsymbolic.

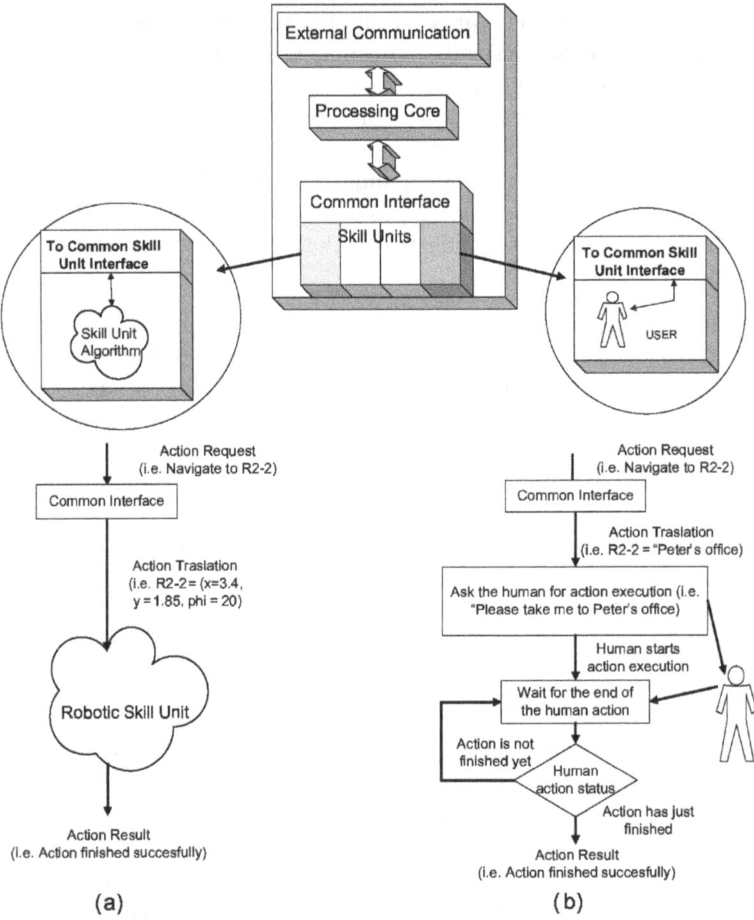



**Fig. 6.2.** The common module structure (CMS). (Center top) All architecture components are designed using this structure. The number of skill units embraced by each module is variable and depends on the human and robot abilities that the CMS provides. (Bottom a) A representation of a robotic skill unit. (Bottom b) A human skill unit. Notice that the human unit requires the natural description of the destination ("Peter office"), which can be automatically labelled in the robot task planning hierarchy as an string like *R2-2*, being needed a translation between both symbols. To perform this action, a robotic unit could only require the geometrical position of the destination

and *events*. The client/server mechanism is a one-to-one communication mechanism that allows modules to request/provide action execution, information, etc., while events are a one-to-many communication mechanism, that is, a signal which is ideally, simultaneously communicated to all modules. In our implementation, events are used to broadcast system alerts, like collision risks or battery failures, to every module of ACHRIN.

Next sections detail the human integration at every tier of the architecture, while describing their modules.


## 6.4 The Deliberative Level

The deliberative layer of ACHRIN is in charge of maintaining and manipulating a symbolic and multi-hierarchical model of the environment. The modules entailed in this layer are: the *ELVIRA framework* module that entails the submodules commented in Chap. 5 (Grounding Agent, Hierarchy Optimizer and Task-Planner), the multi-hierarchical world model that provides mechanisms to access to the internal data, and the *Task Manager*, that manages goal requests to be planned and executed by the robotic system.


### 6.4.1 Symbolic Management

Apart from the use of suitable voice interfaces [150], [154], to approach a high level communication (goal (d) in the introduction), it is needed to endow the robot with a world representation compatible with the human internal representation of space (*cognitive map*).

In our implementation we use a multi-hierarchical world model with two hierarchies[3] (please refer to Fig. 4.7 in Chap. 4), in which the cognitive hierarchy (manually constructed) is devoted to improve the robot interaction with people in a human-like manner, while the other one (automatically generated by *ELVIRA*) is engaged with efficient task planning.

Within the Grounding Agent module (included in ELVIRA), different skill units are devoted to create and manipulate world symbolic information. In our current implementation there is a human skill unit that permits a human to create spatial symbols like distinctive places or rooms in the cognitive hierarchy. To do that she/he must manually guide the vehicle to a certain location where geometrical information, like pose information, is automatically added to the internal model. Topological information is created by the user through the commands (0)-(3) of Fig. 6.3, establishing verbally linguistic labels for topological elements, i.e. "this is a distinctive place called Peter's office". We also consider a robotic skill unit devoted to automatically acquire symbolic information from the environment like rooms and simple objects as explained in Sect. 5.3.2.

These ground spatial symbols, created by both the human and the robot, become the base elements for the automatic construction/tuning of the planning hierarchy 5.4).

---

[3] Robot self-localization is carried out through the cognitive hierarchy, due to the resemblance between grouping places for localization and the human concept of room.

| Id | Human verbal commands | Robot responses |
|---|---|---|
| (0) | This is a distinctive place called <free string> | Ok, place <free string> added. |
| (1) | Group previous places into room <free string> | Ok, room <free string> added. |
| (2) | Open door between <place1> and <place2> | Ok. |
| (3) | Closed door between <place1> and <place2> | Ok. |
| (4) | No alerts | Setting normal state. |
| (5) | Low collision risk | Setting low level alert state. |
| (6) | High collision risk | Setting high level alert state. |
| (7) | Stop | Ok, stopping. |
| (8) | Cancel | Ok, canceling action. |
| (9) | Continue | Ok, continuing the navigation. |
| (10) | Take me to <distinctive place> | Ok.// That place does not exist. |
| (11) | I want to guide you | Ok. |
| (12) | Select another method | Ok. // No available methods, May you help me? |
| | **Robot commands to the human** | **Accepted human responses** |
| (13) | Please can you guide us to <distinctive place> | Yes.// No, I can not. |
| (14) | Can you open the door which is in front of you | Yes. // No, I can not. |
| | **Human acknowledgment information** | **Accepted human responses** |
| (15) | I have guided you to <distinctive place> | Thank you. |
| (16) | I could not reach to <distinctive place> | I can not continue, please select another destination. |
| (17) | I have just opened the door | Thank you. |
| (18) | I could not open the door | I can not continue, please inquire external help. |

**Fig. 6.3.** Human-Robot verbal communication. This table describes the verbally interaction between the user and the robot which has been considered in our experiences. To improve communication we have extended the grammar to accept small variations, i.e. "Let's go to ⟨distinctive place⟩" is recognized as a variant of the command (10)

## Task Planner

Within ACHRIN, the task-planning process deserves especial attention. It is aimed to consider not only robotic skills, but also human physical abilities to solve the requested goals (reported by the Task Manager, see Sect. 6.4.2).

In our current scheme, the *Task Planner* module implements HPWA to efficiently produce plans taking into account the task planning hierarchy from the World Model module and the available set of human and robot abilities from a previously defined planning domain (see Fig. 6.4).

It is relevant to notice that the Task Planner module does not only produce a sequence of actions, but it may also suggest the most efficient method (the skill unit from the functional layer) to perform each one. To do that, we use *Metric-FF* [76] as embedded planner of HPWA since it can produce an optimized plan with respect to a given criteria. In our implementation, plans are optimized with respect to hand-coded cost functions which yield approximations of the execution cost of skill units. For example, function (6.1) yields an approximate cost of a reactive navigation.

$$Cost(Reactive\#1, l1, l2) = k_1 * distance(l1, l2) \tag{6.1}$$

```
(define (domain navigation)
(:requirements :typing :fluents)
(:types location object)
(:const Robot object)
(:predicates
            (at ?obj - object ?loc - location)
            (link ?x ?y - location)
            (open ?obj -object)
            (closed ?obj -object))
(:functions
            (cost-reactive ?l1 ?l2 - location)
            (cost-manually-guided ?l1 ?l2 - location)
            (navigation-cost))
(:action OPEN-MANUALLY-DOOR
      :parameters
      (loc-from - location
      ?door - object)
:precondition
      (and (at Robot ?loc-from)
      (closed ?object))
      (increase navigation-cost 10)
:effect
      (and (not (closed ?object)) (open ?obj)))
(:action MOVE-REACTIVE
:parameters
      (loc-from - location
      ?loc-to - location)
:precondition
      (and (at Robot ?loc-from)
      (link ?loc-from ?loc-to))
:effect
      (and (not (at Robot ?loc-from)) (at Robot ?loc-to)
      (increase navigation-cost (cost-reactive ?loc-from ?loc-to))))
(:action MOVE-MANUALLY-GUIDED
:parameters
      (loc-from - location
      ?loc-to - location)
:precondition
      (and (at Robot ?loc-from)
      (link ?loc-from ?loc-to))
:effect
      (and (not (at Robot ?loc-from)) (at Robot ?loc-to)
      (increase navigation-cost (cost-manually-guided
      ?loc-from ?loc-to)))))
```

**Fig. 6.4.** Example of planning domain. It includes three human-robot abilities, two navigation methods performed by a robotic and a human skill unit respectively, and a manipulation method (open-door) also performed by the human. The parameter **navigation-cost** yields the planning cost involved in navigation tasks

where $l1$ and $l2$ are the origin and destination locations respectively, and $k_1$ is a constant value that measures a certain metric like the time spent by the navigational algorithm to travel each distance unit. The cost of human actions is fixed to a high value to avoid the planning process to select human abilities whereas there are alternative robotic ones.

Thus, through the planning domain described in Fig. 6.4 and the initial situation depicted in Fig. 4.3 in Chap. 4 (considering the door $D3$

closed and all the rest opened), the resultant plan that solves the task *"Go to R2-2"* could be: `MOVE-REACTIVE (H2,WC3)`, `OPEN-MANUALLY-DOOR (D3)`, `MOVE-REACTIVE (D3, R2-2)`, in which the human help is only required to open the closed door.

Notice that a user could not understand a robotic utterance like "Please open D3" in which $D3$ has no special meaning for her/him. Thus, in this case, a proper translation of the symbol $D3$ to the cognitive hierarchy is needed to enable the user to help the robot, that is, to open, for example, "the entrance door of the laboratory" (as explained in Chap. 4).

Human integration at the deliberative level also enables her/him to interactively participate in the planning process. As commented in Chap. 4, the planning process can use the hierarchical arrangement of the internal model to produce plans at a certain intermediate level of abstraction, consulting the user whether such scheme for a plan (an abstract plan) meets her/his wishes. This feature permits humans to accept or reject the proposed plan inquiring a different solution or providing a new one to attain the goal. That is, the human can actively participate at planning stage.

### 6.4.2 Task Manager

The Task Manager module receives requests of tasks to be planned and carried out by the robot. In our particular scheme, only one user can request robot tasks, but in general it could be useful to accept requests from other agents: for example from a remote user that supervise the robot work, or from software applications, i.e. a system procedure that checks the battery level of the robot could request a battery-recharge task.

Skill units of the Task Manager module attend requests from those different "clients"[4]: robotic units should permit applications, or even other machines to request the robot for some operation, while human skill units enable people (in our case only the wheelchair user) to ask the robot for task execution via a voice interface.

## 6.5 The Executive Level

The Execution and Control layer works as an intelligent bridge between the deliberative and the functional layer. It asks modules of the functional layer for the execution of basic actions. During the execution of actions, it also processes the information gathered from sensors to check and avoid risky situations, such as collisions, power failures, etc., deciding the most convenient robot reaction, for instance, stopping the vehicle and waiting for human help.

---

[4] In case of multiple clients requesting tasks to the mobile robot, the processing core of the Task Manager CMS should implement a priority policy between its different skills units.

Human integration plays a relevant role in this layer, since she/he can provide her/his perceptual ability (goal (b) in the introduction) to report danger situations, i.e. "there is a collision risk", as well as to decide what to do when an anomalous circumstance occurs, i.e. "I want to take the motion control" (please, refer again to table of Fig. 6.3 for all available human commands). These features provided by ACHRIN are of special significance for improving robot dependability.

The execution and control layer is composed of two modules: the *Alert System* (Sect. 6.5.1) module which registers external stimuli and the *PLEXAM* module (Sect. 6.5.2) that sequences and controls plans' executions.

### 6.5.1 Alert System

The Alert System module checks for unexpected dangerous situations through its skill units. Robotic skill units can warn the robot about, for example, collisions or low-battery situations through the reading of the robot sensors. In addition, the wheelchair user is integrated into the Alert System through human skill units which incorporate the capability to inform about risky situations using verbal commands like "there is a collision risk" or "the battery level is low".

The integration of the human into the alert system of ACHRIN improves the reliable performance of the robot, since she/he can report danger situations not detected by robot sensors. Besides, the human can also predict future risky situation based on her/his knowledge about the environment, and thus, for example, she/he can notice about a collision risk when the vehicle is close to enter a room which is probably crowded.

The Alert System distinguishes different risk levels to adequate the behavior of the vehicle operation (see Fig. 6.5). Thus, for the case of collision alerts[5] we have defined the following four alert levels based on both human indications and fix thresholds for readings of the laser scanner of our robot[6]:

- *Normal.* No obstacles have been detected, and thus the system works without any danger.
- *Minimal Risk.* An obstacle has been detected at a safety distance, and thus a minimal risk is considered. This low-level risk may produce variations on the execution parameters of the current robot action, like reducing the vehicle velocity.
- *Maximum Risk.* A critical situation has been reported by a robotic or human skill unit. In this case, the module that executes the current robot

---

[5] A similar alert categorization can be done for other situations like low-level battery charge.

[6] In our current implementation the sensor-based alerts are triggered based on fixed thresholds, but a more flexible mechanism can be implemented to adapt the vehicle behaviour to the human preferences, learning such a threshold from experience [13].

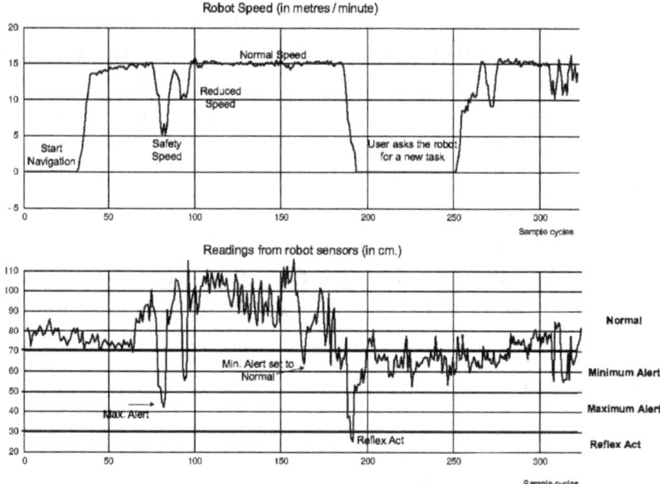


**Fig. 6.5.** Evolution of collision alert levels over time in a real experiment: (Top) Speed of the robot during part of our experiments (1 sample cycle=0.5 sec.); (Bottom) Minimum distance measured by robot sensors to the closest obstacle. Alerts level threshold is marked with a thick line. Notice that the user can ignore, for example, a minimum alert by setting the alert level to a normal state (around cycle 160)

action must modify its parameters to prevent any danger, i.e. reducing drastically the vehicle speed.

- *Reflex Act.* A reflex act is the most critical level. When an obstacle is detected too close, a reflex act is triggered, stopping the vehicle to avoid any possible damage. The user must recover this situation by requesting a new task.

The user plays a dominant role in the Alert System module since alerts reported by her/him are prioritized. Thus, the user can ignore, for instance, a minimal risk from a collision skill unit that uses the readings of a laser rangefinder sensor, setting the alert level to normal[7] (see Fig. 6.5). This feature permits humans to modulate the behavior of the robot to her/his wills, i.e. when the user, intentionally, wants to closely approach a wall.

Chart of Fig. 6.5 shows the evolution of the collision alert levels in part of our experiments in which three different situations can be observed: (1) A maximum alert is detected near cycle 80, decreasing the robot speed to a safety value (5 m/min). (2) Minimal risk alerts (around cycle 160) are ignored by the user who establishes a normal state (vehicle speed to 15 m/min).

---

[7] When the user ignores an alert caused by a certain sensor, its readings are ignored during a certain period of time (set to 5 secs. in our implementation), only attending human alerts. After that period of time, the system automatically considers again any possible alert from any source.

(3) Finally, a collision reflex (before cycle 200) is detected causing the detention of the vehicle and waiting for a new task.

Other robot architectures also include alert mechanisms (see for example [31], [46]) but they do not incorporate the user capability for consciously ignoring certain alerts from the environment, which is of much interest in cluttered scenarios.

### 6.5.2 Plan Executor and Alert Manager (PLEXAM)

This module sequences plans' actions and controls their execution based on the information gathered by the alert system. During the execution of a particular action, PLEXAM registers system alerts according to their severity and communicates them (if any) to the functional layer in order to accordingly tune their performance (see Figs. 6.5 and 6.7).

PLEXAM is also in charge of detecting malfunctions of skill units. For instance, a failure in the execution of a navigation skill is considered when it has not responded (reporting neither success nor failure) after a certain period of time. In that cases, PLEXAM is responsible for cancelling the current execution and deciding the next skill unit to finish the action.

The human is also integrated into this module by a human skill unit that enables the user to report execution failures (goal (c) in the introduction) deciding the selection of another skill unit to finish the action given the current environmental conditions. In the case of navigation failures, for example, the user can decide to take the control of the vehicle through the manually-guiding method. Such a participation of the human in the robot navigation is restricted, as any other functional component of ACHRIN, by the safety issues imposed by the Alert System. Thus, when the user guides the vehicle, PLEXAM can stop it due to a collision alert, ignoring the user motion inputs and choosing the next available skill unit to accomplish the action. In the case of failure of all available skill units (including the human ones), PLEXAM cancels the execution of the current plan, inquiring the user for a new achievable goal.

In the absence of failures, when the functional layer reports the successfully execution of the requested action, PLEXAM continues the execution of the next action (if any) of the plan. Figure 6.6 details the PLEXAM work through a state graph.

## 6.6 The Functional Level

The Functional (and lowest) layer of the architecture contains a set of modules that physically materialize the abilities of the human-robot system. This layer must contain at least a functional module to perform navigation; however, additional modules can be also considered to perform other kind of actions such as manipulation, inspection, social interaction, etc. Besides, each module

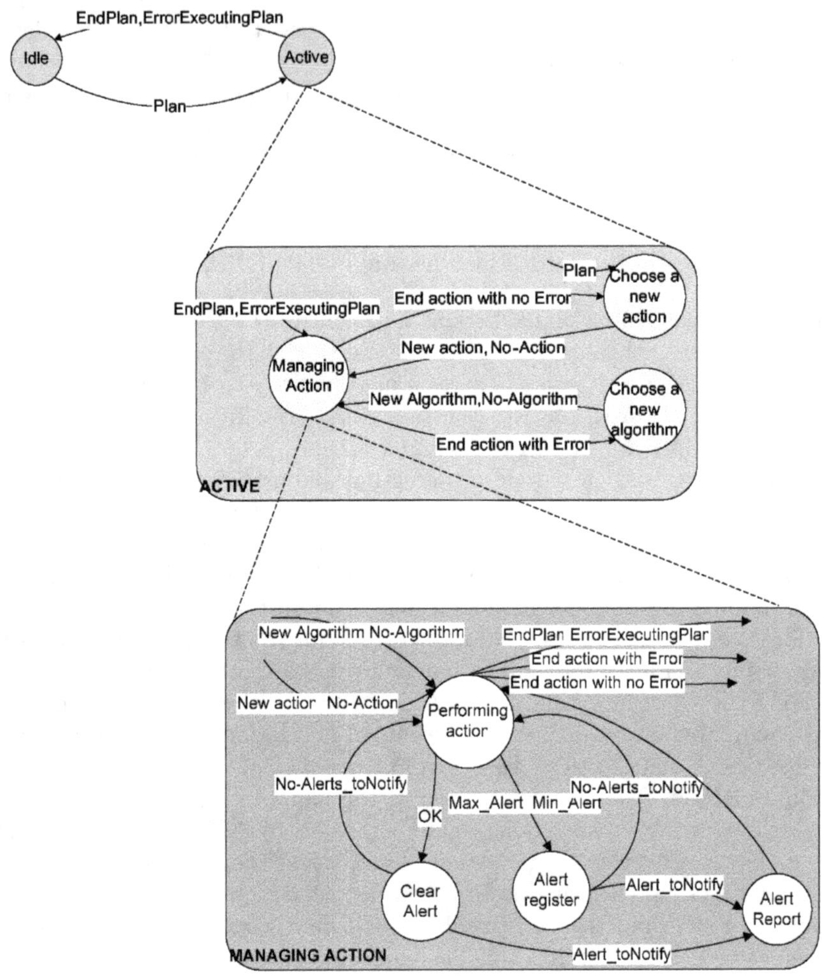


**Fig. 6.6.** PLEXAM state graph. This module sequences actions while maintaining a list of all alerts registered in the system. It notifies alerts to functional modules at the same time that it waits for their results. Based on such results, PLEXAM selects a new action or a new algorithm (skill unit to perform the next action)

can entail a number of skill units to perform an action, i.e. navigation, in different ways (reactive, tracking a computed path, or manually guided).

In this layer, the human is integrated into the robotic system augmenting and/or improving the robot capabilities (goal (a) in the introduction), i.e. an user can help the vehicle to manipulate objects (open a door) or she/he can recover the robot from a navigation failure, manually guiding the vehicle as shown in Sect. 6.7.3.

The processing core of functional modules takes care of alerts reported by PLEXAM to establish the execution parameters of skill units, i.e. the

**Speed parameter (meter/minute)**

| Navigation skill Unit | Normal | Minimal Risk | Maximum Risk |
|---|---|---|---|
| Manually guided | 12 | 10 | 5 |
| Reactive | 15 | 10 | 5 |
| Tracked | 10 (from 9:00 to 12:00am) <br> 20 (the rest of the day) | 8 | 4 |

**Fig. 6.7.** Example of the variation of the speed parameter for navigation skill units under alerts occurrence

maximum speed of a reactive algorithm in a normal situation (no alerts are reported) is set to 15 metres/minutes and to 10 when a minimal risk is reported (see Fig. 6.5). Such values can be either fixed and hand-coded for every skill unit or variable depending, for instance, on a particular period of time (see table of Fig. 6.7).

The execution of skill (human or robotic) units from the functional layer may finish due to three different causes:

- The unit successfully executes its action. This is the normal ending status of the skill units which is communicated to PLEXAM to launch the next action (if any) of the plan.
- The execution is cancelled by PLEXAM. It can cancel the execution of a skill because of a detected failure.
- The skill unit execution finishes without achieving its goal due to an unforeseen circumstance or a failure.

## 6.7 Experiences on the SENA Robotic Wheelchair

ACHRIN has been largely tested within human, large scenarios on the robotic wheelchair SENA. More precisely, experiences were carried out at the Computer Engineering Building in the University of Málaga. This section firstly describes the hardware details of SENA. Section 6.7.2 presents some implementation details of the software developed for SENA. Finally, Sect. 6.7.3 describes some reals experiences carried out on SENA.

### 6.7.1 The SENA Robotic Wheelchair

*SENA* is a robotic wheelchair (see Fig. 6.8) based on a commercial powered wheelchair (Sunrise Powertec F40 [107]) that has been equipped with several sensors to reliably perform high-level tasks in office-like environments. This mobile robot is the result of years of research at the System Engineering and Automation Department at the University of Málaga, in which a large number of researchers and undergraduate students have participated.

When designing SENA, two main aspects were considered: on the one hand, to keep unalterable the original elements of the electric wheelchair as

**Fig. 6.8.** The robotic wheelchair SENA. It is based on a commercial electric wheelchair which has been endowed with several sensors (infrared, laser, and a CCD camera). Wheelchair motors as well as sensors are managed by a microcontroller communicated to a laptop computer via USB

much as possible, while, on the other hand, strategically place additional elements, like sensors, to reliably perform their tasks without being a nuisance to the user. Following these design guidelines, SENA uses the original controller and motors, as well as the joystick which can be used by the user to move the vehicle.

SENA is endowed with a complete set of sensors to capture environmental data as well as proprioceptive information:

- A 180 radial laser scanner (PLS) placed in front of the wheelchair, mounted on a retractile mechanism between the user legs for avoiding any nuisance

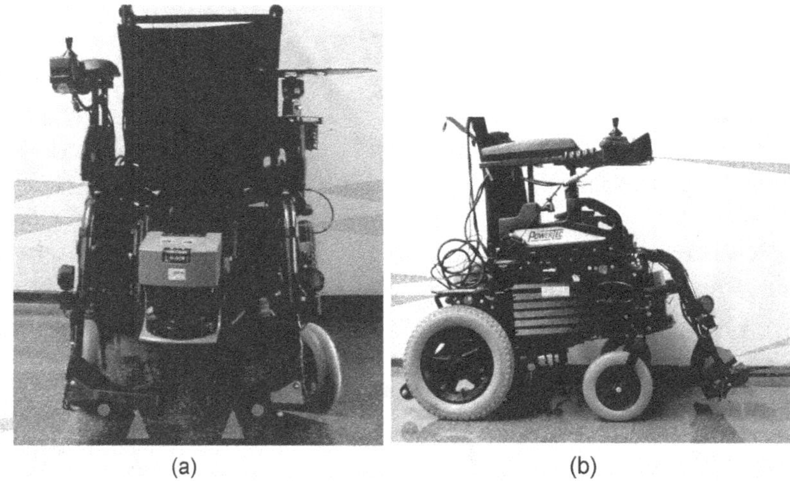



**Fig. 6.9.** Location detail of the infrared sensors of SENA. A ring of twelve infrared sensors are placed around the vehicle. Two of them are located underside (one per footrest) to detect portholes or stairwells

to her/him. It is used to check for mobile obstacles, for environment map construction [66], and for robot localization [124].

- Two ultrasonic rotating sensors also located in front of SENA. Each one is mounted on a servo which enables it to scan a range of 180 detecting transparent or narrow objects which may not be properly captured by laser sensors.
- A ring of twelve infrared sensors placed around SENA to detect close obstacles (see Fig. 6.9). Two of them are located underside of SENA to check for portholes, kerbs, stairwells, etc. Other two infrared sensors are located in the backside of SENA to avoid possible obstacles when it moves backwards.
- A CCD camera situated on a high position and mounted on a pan-tilt unit is used to localize SENA [111].
- Two encoders connected to the motors' axis to estimate the odometric position of the vehicle.

Sensorial (and also motor) management is performed by a microcontroller connected to a laptop computer via USB (see Fig. 6.10). This microcontroller serves as an interface between the laptop and different devices and sensors. Apart from the sensorial system described above, SENA also accounts for a voice system consisting of a bluetooth headset that permits the user to command the wheelchair as well as a pair of small speakers

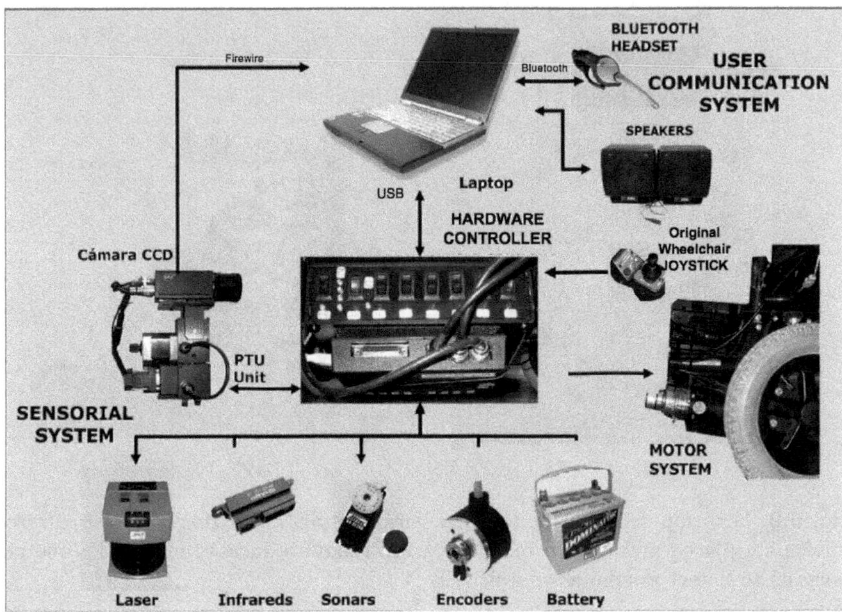


**Fig. 6.10.** General scheme of the hardware of SENA. High-level control algorithms (i.e., modules of ACHRIN) are executed in the laptop, while low-level control of SENA's devices are managed by a microcontroller connected to the laptop via USB

## 6.7.2 Software Details

For implementing our software, we have used a general framework called *the BABEL development system*[48]. Robotic applications, like the one described in this chapter, are usually framed into a long-term project that involves many researchers in different areas of expertise. Moreover, heterogeneous hardware components like computers, microcontrollers, sensor and actuator devices, and software platforms, i.e. operating systems, communication middlewares, programming languages, etc., are required in this kind of applications.

The Babel development system copes with the main phases of the application lifecycle (design, implementation, testing, and maintenance) when unavoidable heterogeneity conditions are present. It entails a number of tools aimed to facilitate the work of programmers in order to speed up the development of robotic applications. For instance, the *BABEL Module Designer* automatically generates the needed code to distribute software developed in a certain programming language, i.e. C++, through a particular communication middleware like CORBA [70]. Another relevant tools within the BABEL system are the the *BABEL Execution Manager (EM)*, the *BABEL Development Site (DS)*, and the *BABEL Debugger (D)*. The *EM* facilitates the execution of the different components of a distributed application; *DS* serves as a software repository accessible from internet (*http://babel.isa.uma.es/babel/*), and $D$ is a tool that recovers information produced during execution.

In the experiments carried out in our work (see Sect. 6.7.3) we have intensively used the tools entailed in the Babel development system. Our experience has largely proved the benefits of their use.

### 6.7.3 Experiences on SENA

The robotic wheelchair SENA has served as a test-bed for the work presented along this book. It has been used to physically demonstrate the performance of the robotic architecture ACHRIN, which has been entirely developed under the Babel development system.

When demonstrating the suitability of a robotic architecture to a particular application we find that it is not easily quantifiable. We have tested ACHRIN by performing several real experiences on SENA, within a real environment (the research building of the Computer Engineering Faculty of the University of Málaga) and with different people who have confirmed their satisfaction while using the vehicle (see Fig. 6.11).

In our tests we have use the SENA navigation abilities through a reactive algorithm. Sensors available in our experiences for navigation and obstacle detection/avoidance are a laser rangefinder and a ring of eight infrared sensors. From the human part, we have considered that the user is enabled to improve the navigation of the robot by manually guiding the vehicle and to extend its abilities by opening doors. She/he can also alert the system from collision alerts as well as detect system failures.

For a natural human-robot interaction, commercial voice recognition/ generation software [150], [154] has been used. The set of considered commands and their possible responses have been enumerated in Fig. 6.3 (in the following, commands' IDs will be used to describe our tests).

The first part of our tests is endowing SENA with a symbolic model of the workspace used to plan and execute tasks (see [47] for a detailed description).

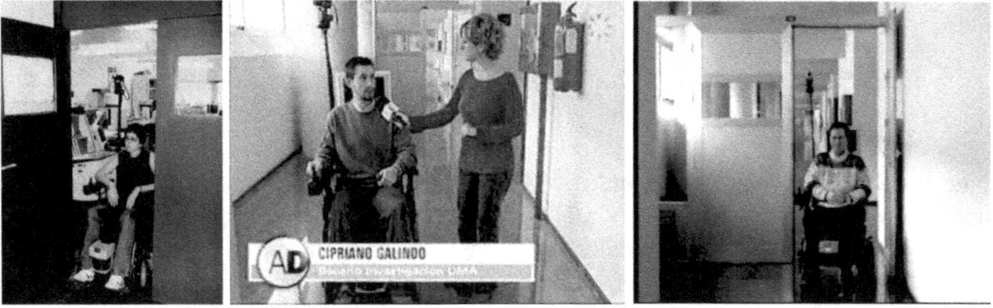


**Fig. 6.11.** Experiments with the SENA robotic wheelchair. These images are snapshots of videos (some of them appeared in life tv shows), which can be found at [65, 64]

This model is created by the Grounding Agent within the deliberative layer of
ACHRIN through a human skill unit that responds to commands (0)–(3), and
a robotic skill unit that automatically creates ground symbols. Geometrical
information (like the robot pose) needed for posterior navigation is automatically attached to vertexes and edges. Figure 6.12 depicts an abstract level of
the internal model managed by SENA as well as the spatial hierarchy used in
our experiences.

Once a model of part of the environment is available (for example, a couple
of rooms and a corridor), the user can select a destination through a human
skill unit within the Task Manager module that attends to command (10), for
instance "Take me to the *Copier*". Using the information stored in the World
Model module (Fig. 6.12) and the planning domain of Fig. 6.4, the resultant
plan, $P$, yielded by the Task Planner module is shown in Fig. 6.13.

In the generation of plan $P$, human help is only considered when there is
not an available robot ability to achieve the goal, i.e., open a door. During
the plan execution (see Fig. 6.15 a), in absence of failures on the robotic unit
that performs navigation, PLEXAM only inquires the human help (through

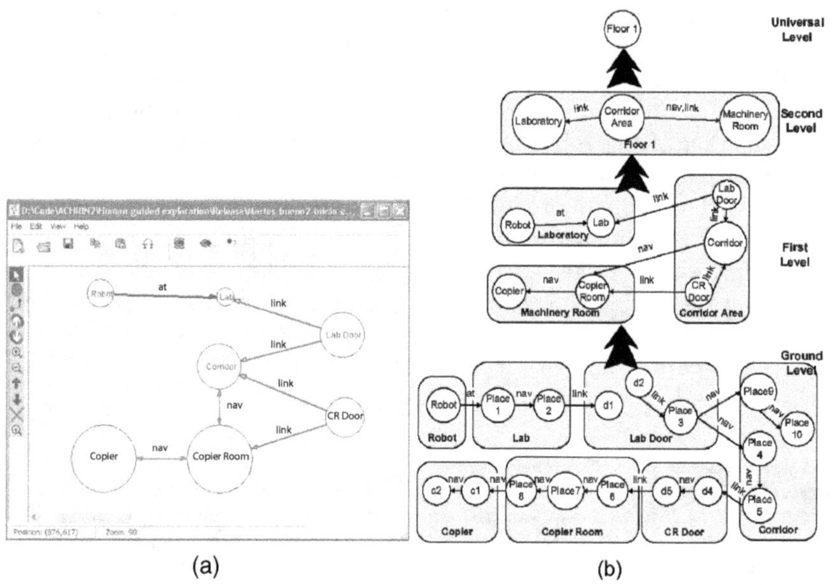

(a)            (b)

**Fig. 6.12.** Hierarchical representation of part of the environment of SENA: (a) The
internal world model with linguistic labels is shown through a graphical interface
(only the first level). Edges indicate relations between spatial symbols, like navigability. The state of doors (opened or closed) is deduced from vertex connectivity,
i.e., "Lab door" is closed since there are not navigability edges between "Lab" and
"Corridor"; (b) Spatial hierarchy considered in our experiments. Ground and first
level are created by the human-robot system, while the others have been include for
completeness

| Plan P |
| --- |
| MOVE-REACTIVE *Lab, Lab-door* |
| OPEN-MANUALLY-DOOR *Lab-door* |
| MOVE-REACTIVE *Lab-door, Corridor* |
| MOVE-REACTIVE *Corridor, Copier-Room-door* |
| MOVE-REACTIVE *Copier-Room-door, Copier* |

**Fig. 6.13.** Sequence of skills to go from *Lab* to the *Copier*

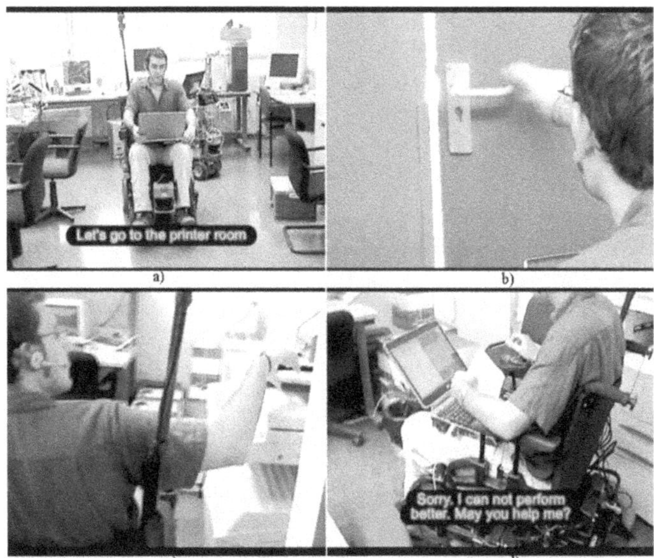


**Fig. 6.14.** Execution of navigation tasks: (a) The human asks the robot for the execution of a task; (b) The robotic system requires the human help to open a door; (c) The vehicle reaches the destination: the copier-room; (d) The user recover the robot after a navigation failure

command (14)) after the success of the first action (MOVE-REACTIVE *Lab, Lab-door*). When the user reports the success of the entrusted action (human acknowledgement (17))[8], the execution of $P$ continues autonomously until the destination is achieved. Figure 6.14 shows some pictures taken during the execution of plan P.

Our experiences assume that the information stored in the internal model is coherent with the real environment and that there are not external agents modifying the robot workspace. Thus, during planning no attention has been paid to model incoherences like opening a door which is just opened or traversing a closed door. However, failures produced for such inconsistencies are

---

[8] Note that opening a door while sitting on a wheeled chair may turn into an arduous task. In some cases in which the user could not open the door she/he could ask surrounded people for help.

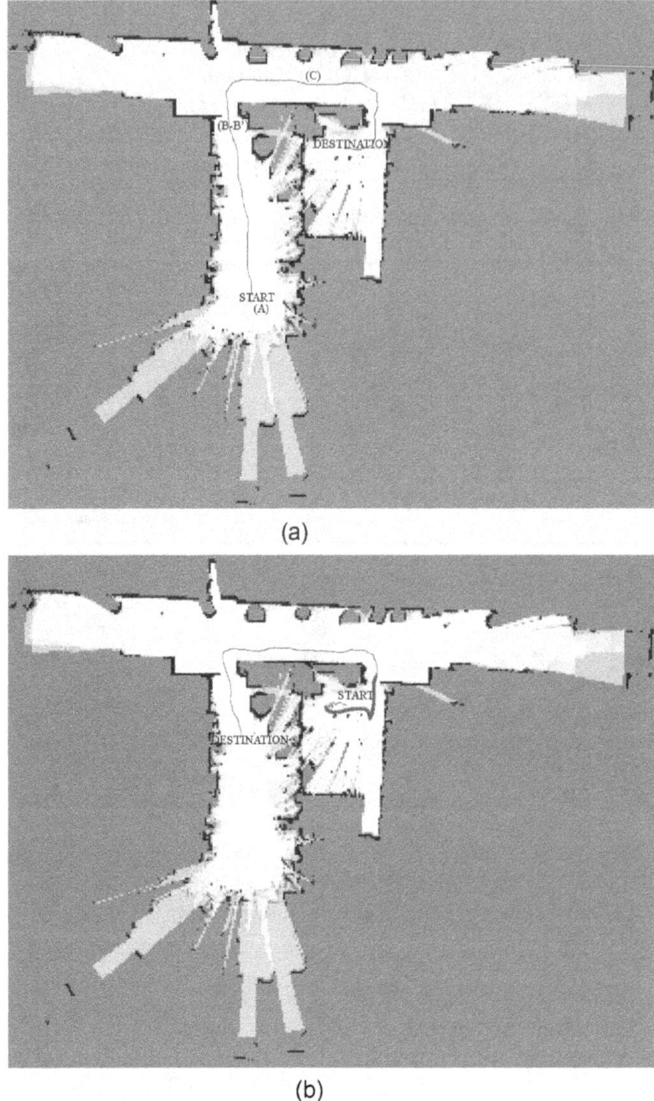



**Fig. 6.15.** Occupancy gridmaps from our test scenario: (a) The route marked in the map corresponds to the navigation of the vehicle during the execution of the plan $P$; (b) Path followed during the execution of $P^*$. The thicker section of the path corresponds to the manually guidance

correctly managed by human integration. In the latter situation, for example, the integrity of the system is ensured by the alert system which would report a collision alert stopping the vehicle while the user should modify the internal model of the system through command (3).

In other experiences, we have also tested human-robot integration at the intermediate level of ACHRIN to control the execution of plans. Thus, after the execution of plan $P$, the user asked SENA to go back to the laboratory (command (10)). In this case the resultant plan $P^*$ is similar to $P$ but in a reverse order and without the "open door" action, since it is supposed to be opened after the execution of plan $P$. Plan $P^*$ only contains navigational actions, thus the Task Planner module does not consider human participation to solve the task. The execution of $P^*$ starts normally with the PLEXAM call to the robotic skill unit that performs the action MOVE-REACTIVE (*Copier, Copier-Room-door*). In the test scenario, our reactive algorithm usually fails when performing this action due to the lack of space for manoeuvering (see Fig. 6.15 b). In this case, users normally reports a navigation malfunction of the robotic unit through a human skill unit within PLEXAM, cancelling the vehicle navigation (command (8)) and inquiring the selection of an alternative method (command (12)). The human help is required by PLEXAM via command (13), "Can you take me to the *Copier-Room-door?*". Once the user concludes the action and reports the success of the manually guidance of the vehicle to the destination, the execution of the plan is resumed. Figure 6.16 shows the flow of information between modules during the execution of plan $P^*$.

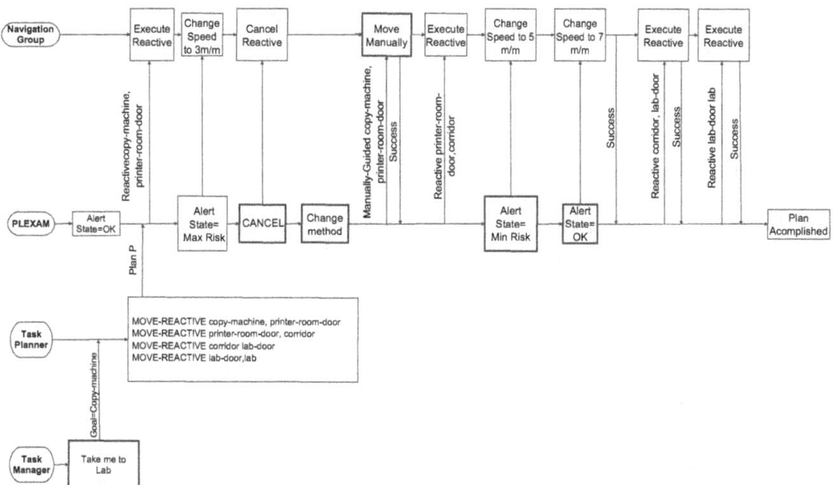

**Fig. 6.16.** Scheme of some modules' communication during the execution of a plan. The flow of information is from bottom to top and from left to right. Thick boxes represents human skill units

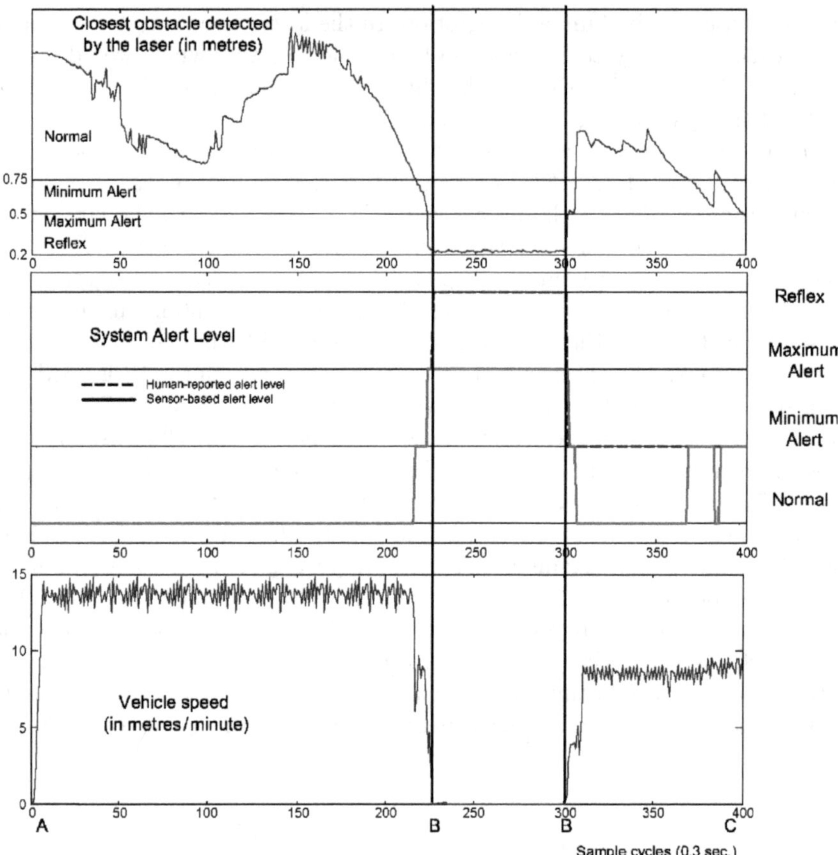



**Fig. 6.17.** Collision alerts test: (Top) Distance to the closest obstacle during part of the wheelchair navigation measured by the laser rangefinder; (Middle) System alert level along the experience. Notice how the user can set a certain alert level more restrictive than the sensor-based alert; (Bottom) The vehicle speed during the experience. Speeds depends on the system alert level (15, 10, and 5 m/min. for normal, minimum and maximum levels, respectively). In the interval $B$-$B'$, in which the user triggered a reflex act, the speed is set to 0

Notice that when the human help is required to accomplish an action she/he can refuse the order or may not properly finished it, for example, she/he may not be able to open a closed door. In that cases, since human help is only required as a last resort, the system stops, waiting for external help (from a surrounded person) or another achievable goal commanded by the user.

System collision alerts were also tested in our real experiences. During the navigation of the wheelchair, i.e. executing plans like $P$, the alert system continuously checks for collision risks from two sources: the rangefinder sensors (laser scanner and infrared ring) mounted on SENA and the alert messages reported by the user (through commands (5)–(7)). Figure 6.17-top shows the distance yielded by the rangefinder sensors to the closest obstacle during the navigation from point $A$ to $C$ (see Fig. 6.15. Based on this information and the defined alert thresholds (0.2 m. for reflex, 0.5 m. for maximum alert, and 0.75 m. for minimum alert), Fig. 6.17-middle shows the evolution of the system alert level during a section of an experiment along a clutter environment. In this Fig. we distinguish the alert source: human or sensor-based alert. Human alerts prevail over sensor-based ones, and thus, in this example, between points $B$ and $B'$ (around cycles 225-300) the user sets the alert level to *Reflex* near the open door, albeit the rangefinder readings report a *Maximum alert*. In the same manner, from point $B'$ to the destination (point $C$) the alert level is set to *Minimum alert* by the user in spite of the rangefinder information. Finally, Fig. (6.17-bottom) shows the robot speed during the navigation and its adaptation to the current alert level.

## 6.8 Conclusions

This chapter has proposed the integration of humans into robotic systems as the best way to extend/augment mobile robot performance in real scenarios. This simple, but effective, solution enables those robots that closely work with people to overcome their lack of autonomy within human environments.

Human integration is achieved at all levels of the architecture by considering human abilities as software algorithms which are integrated into all components of ACHRIN. To this aim, the use of our multi-hierarchical and symbolic model has played a significant role, providing efficiency in planning as well as a proper human-robot communication. The implementation of ACHRIN on a real robot like SENA, has permitted us to test its suitability to this kind of assistant robotic applications.

# 7

# Conclusions

*Per aspera ad astra!*
Latin proverb (To the stars through difficulties)

This book has addressed the problem of endowing an agent with an efficient and coherent symbolic representation of its workspace. In spite of the problems arising from symbolic modeling, i.e. symbol genesis, an agent that accounts for a symbolic world representation exhibits, among others, two significant advantages:

- Capability for planning, that is, capability to devise a sequence of actions that, in theory, should yield the desired objective.
- Capability for communicating to other agents who share a common set of symbols.

These two characteristics are clearly present in humans, and thus, they should be considered in the development of agents intended to operate in human environments, like our robotic taxi driver.

Along this book we have formalized and utilized a multi-hierarchical symbolic model in order to provide a mobile robot with the commented features. Moreover, we have taken into account the computational complexity of planning tasks in large domains by implementing planning algorithms that exploit the hierarchical arrangement of the symbolic information. Such an arrangement of the model is continuously tuned (optimized) in order to cope efficiently with changes.

Regarding human-robot communication, given that we permit humans to create symbols in the internal world model of the robot, they (human and robot) can communicate to each another using the same terms.

All these characteristics have been tested on a assistant robotic application, a robotic wheelchair that has played the role of "taxi driver".

So, are we now closer to the *Robotic Future*? From the deliberative point of view, the mechanisms presented in this book provide a steady step towards intelligence and autonomy of mobile robots. Nevertheless, we have not tackled

C. Galindo et al.: *A Multi-Hierarchical Symbolic Model of the Environment for Improving Mobile Robot Operation*, Studies in Computational Intelligence (SCI) **68**, 143–144 (2007)
`www.springerlink.com`                                      © Springer-Verlag Berlin Heidelberg 2007

functional issues, such as uncertainty when executing plans, sensor reliability, fault tolerance, and robustness[1] which are unavoidable in uncontrolled environments.

Thus, what are the next steps? The authors are aware of the complexity of the way to the development of fully autonomous and intelligent (or at least pseudo-intelligent) robots, as well as of the limitations of the work presented in this book. Lines that deserve a deeper study and research include (but are not limited to):

- Improving the automatic grounding mechanisms for reliably acquiring symbolic information from uncontrolled environments, relieving humans from this task.
- Permitting agents to automatically interchange their anchored symbols. This could permit an agent to expand its internal world model with the information (experience) of other agents.
- To explore the inclusion of semantic information into the model (in the line of [60]) to provide the robot with extra information about its environment. For example, in the case of the robotic taxi driver, it could use semantic information to infer that a certain street is usually crowded from 12:00 to 14:00 since it is close to a shopping center, and thus, he could avoid cross it.
- To study the needed mechanisms to provide the robot with robustness, and fault tolerance within real and non-controlled scenarios, for instance, to modify a plan in order to face unexpected situations, i.e. when a door, that is initially supposed to be opened, is in fact locked when the robot approach it.

---

[1] The design of *ACHRIN* considers some of this issues through different replicas of functional components, but this topic has not been explicitly addressed.

# A

# Mathematical Demonstrations for the Formalization of the Graph Category

This appendix provides some mathematical demonstrations related to the formalization of the category of graphs with abstraction and refinement given in Chap. 2.

## A.1 Composition of Abstractions in *AGraph* is an Abstraction

This section demonstrates what is stated in (2.12), that is, that $A_2 \Diamond A_1$ satisfies constraints for abstractions of graphs. The contraints that function $\Diamond$ must satisfy for being considered an abstraction are (2.4-2.8), that we repeat here for convenience:

$$\nu : V^{(G)} \rightarrow V^{(H)} \ is \ a \ partial \ function. \tag{A.1}$$

$$\varepsilon : E^{(G)} \rightarrow E^{(H)} \ is \ a \ partial \ function. \tag{A.2}$$

$$\forall z \in E^{(G)}, \ def(\varepsilon(z)) \Rightarrow [def(\nu(ini^{(G)}(z))) \wedge def(\nu(ter^{(G)}(z)))] \tag{A.3}$$

$$\forall z \in E^{(G)}, \ def(\varepsilon(z)) \Rightarrow [\nu(ini^{(G)}(z)) \neq \nu(ter^{(G)}(z))] \tag{A.4}$$

$$\forall z \in E^{(G)}, \ def(\varepsilon(z)) \Rightarrow \begin{bmatrix} \nu(ini^{(G)}(z)) = ini^{(H)}(\varepsilon(z)) \wedge \\ \nu(ter^{(G)}(z)) = ter^{(H)}(\varepsilon(z)) \end{bmatrix} \tag{A.5}$$

*Demonstration of (A.1) and (A.2)*

Constraints (A.1) and (A.2) are satisfied by the definition of function $\Diamond$ given in (2.12). Since $\nu$ is partial (by (2.4)), then $\nu^{(A_2)}(\nu^{(A_1)}(a))$ must also be partial. Since $\varepsilon$ is partial (by (2.5)), then $\varepsilon^{(A_2)}(\varepsilon^{(A_1)}(z))$ must also.

$\blacksquare$

*Demonstration of (A.3)*

For demonstrating (A.3), we must demonstrate that:

$$\forall z \in E^{(G^{(A_1)})}, def(\varepsilon_o(z)) \Rightarrow [def(\nu_o(ini^{(G^{(A_1)})}(z))) \wedge def(\nu_o(ter^{(G^{(A_1)})}(z)))] \tag{A.6}$$

By (2.12) we can deduce that:

$$\forall z \in E^{(G^{(A_1)})}, def(\varepsilon_o(z)) \Rightarrow def(\varepsilon^{(A_1)}(z)) \wedge def(\varepsilon^{(A_2)}(\varepsilon^{(A_1)}(z))) \tag{A.7}$$

And using (2.6) and (2.8) on (A.7), we can go on with the implication:

$$\forall z \in E^{(G^{(A_1)})}, def(\varepsilon_o(z)) \Rightarrow$$
$$def(\varepsilon^{(A_1)}(z)) \Rightarrow$$
$$\begin{bmatrix} def(\nu^{(A_1)}](ini^{(G^{(A_1)})}(z)) \wedge \\ def(\nu^{(A_1)}](ter^{(G^{(A_1)})}(z)) \end{bmatrix} \wedge \tag{A.8}$$
$$\begin{bmatrix} \nu^{(A_1)}(ini^{(G^{(A_1)})}(z)) = ini^{(H^{(A_1)})}(\varepsilon^{(A_1)}(z)) \wedge \\ \nu^{(A_1)}(ter^{(G^{(A_1)})}(z)) = ter^{(H^{(A_1)})}(\varepsilon^{(A_1)}(z)) \end{bmatrix}$$

$$\forall z \in E^{(G^{(A_1)})}, def(\varepsilon_o(z)) \Rightarrow def(\varepsilon^{(A_2)}(\varepsilon^{(A_1)}(z))) \Rightarrow$$
$$\begin{bmatrix} def(\nu^{(A_2)}(ini^{(G^{(A_2)})}(\varepsilon^{(A_1)}(z)))) \wedge \\ def(\nu^{(A_2)}(ter^{(G^{(A_2)})}(\varepsilon^{(A_1)}(z)))) \end{bmatrix} \tag{A.9}$$

But since $H^{(A_1)} = G^{(A_2)}$, (A.9) can be rewritten:

$$\forall z \in E^{(G^{(A_1)})}, def(\varepsilon_o(z)) \Rightarrow def(\varepsilon^{(A_2)}(\varepsilon^{(A_1)}(z))) \Rightarrow$$
$$\begin{bmatrix} def(\nu^{(A_2)}(ini^{(H^{(A_1)})}(\varepsilon^{(A_1)}(z)))) \wedge \\ def(\nu^{(A_2)}(ter^{(H^{(A_1)})}(\varepsilon^{(A_1)}(z)))) \end{bmatrix} \tag{A.10}$$

And using (A.8),

$$\forall z \in E^{(G^{(A_1)})}, def(\varepsilon_o(z)) \Rightarrow def(\varepsilon^{(A_2)}(\varepsilon^{(A_1)}(z))) \Rightarrow$$
$$\begin{bmatrix} def(\nu^{(A_2)}(\nu^{(A_1)}(ini^{(G^{(A_1)})}(z)))) \wedge \\ def(\nu^{(A_2)}(\nu^{(A_1)}(ter^{(G^{(A_1)})}(z)))) \end{bmatrix} \tag{A.11}$$

Since by (2.12), $\forall a \in V^{(G^{(A_1)})}, \nu_o(a) = \nu^{(A_2)}(\nu^{(A_1)}(a))$, (A.11) is in fact:

$$\forall z \in E^{(G^{(A_1)})}, def(\varepsilon_o(z)) \Rightarrow def(\varepsilon^{(A_2)}(\varepsilon^{(A_1)}(z))) \Rightarrow$$
$$\begin{bmatrix} def(\nu_o(ini^{(G^{(A_1)})}(z))) \wedge \\ def(\nu_o(ter^{(G^{(A_1)})}(z))) \end{bmatrix} \tag{A.12}$$

which is what we wanted to demonstrate (A.6).

∎

*Demonstration of (A.4)*

For demonstrating that (A.4) holds, we must show that:

$$\forall z \in E^{(G^{(A_1)})}, def(\varepsilon_\circ(z)) \Rightarrow \nu_\circ(ini^{(G^{(A_1)})}(z)) \neq \nu_\circ(ter^{(G^{(A_1)})}(z)) \quad (A.13)$$

which, using the definition of $\nu_\circ$ and $\varepsilon_\circ$ in (2.12), can be rewritten:

$$\forall z \in E^{(G^{(A_1)})}, def(\varepsilon^{(A_2)}(\varepsilon^{(A_1)}(z))) \Rightarrow$$
$$\left[ \nu^{(A_2)}(\nu^{(A_1)}(ini^{(G^{(A_1)})}(z))) \neq \nu^{(A_2)}(\nu^{(A_1)}(ter^{(G^{(A_1)})}(z))) \right] \quad (A.14)$$

But from (2.7) we can deduce that:

$$\forall z \in E^{(G^{(A_1)})}, def(\varepsilon^{(A_2)}(\varepsilon^{(A_1)}(z))) \Rightarrow$$
$$\left[ \nu^{(A_2)}(ini^{(G^{(A_2)})}(\varepsilon^{(A_1)}(z))) \neq \nu^{(A_2)}(ter^{(G^{(A_2)})}(\varepsilon^{(A_1)}(z))) \right] \quad (A.15)$$

Since $H^{(A_1)} = G^{(A_2)}$, this can be rewritten:

$$\forall z \in E^{(G^{(A_1)})}, def(\varepsilon^{(A_2)}(\varepsilon^{(A_1)}(z))) \Rightarrow$$
$$\left[ \nu^{(A_2)}(ini^{(H^{(A_1)})}(\varepsilon^{(A_1)}(z))) \neq \nu^{(A_2)}(ter^{(H^{(A_1)})}(\varepsilon^{(A_1)}(z))) \right] \quad (A.16)$$

And now we can use (2.8) to do:

$$\forall z \in E^{(G^{(A_1)})}, def(\varepsilon^{(A_2)}(\varepsilon^{(A_1)}(z))) \Rightarrow$$
$$\left[ \nu^{(A_2)}(\nu^{(A_1)}(ini^{(G^{(A_1)})}(z))) \neq \nu^{(A_2)}(\nu^{(A_1)}(ter^{(G^{(A_1)})}(z))) \right] \quad (A.17)$$

which is (A.14), what we wanted to demonstrate.

∎

*Demonstration of (A.5)*

Now we need to demonstrate (A.5) for the composition of abstractions, which is:

$$\forall z \in E^{(G^{(A_1)})}, def(\varepsilon_\circ(z)) \Rightarrow \left[ \begin{array}{l} \nu_\circ(ini^{(G^{(A_1)})}(z)) = ini^{(H^{(A_2)})}(\varepsilon_\circ(z)) \wedge \\ \nu_\circ(ter^{(G^{(A_1)})}(z)) = ter^{(H^{(A_2)})}(\varepsilon_\circ(z)) \end{array} \right]$$
$$(A.18)$$

Substituting $\varepsilon_o$ and $\nu_o$ by their definitions in (2.12):

$$\forall z \in E^{(G^{(A_1)})}, def(\varepsilon^{(A_2)}(\varepsilon^{(A_1)}(z))) \Rightarrow$$
$$\left[ \begin{array}{l} \nu^{(A_2)}(\nu^{(A_1)}(ini^{(G^{(A_1)})}(z)) = ini^{(H^{(A_2)})}(\varepsilon^{(A_2)}(\varepsilon^{(A_1)}(z)))\wedge \\ \nu^{(A_2)}(\nu^{(A_1)}(ter^{(G^{(A_1)})}(z)) = ter^{(H^{(A_2)})}(\varepsilon^{(A_2)}(\varepsilon^{(A_1)}(z))) \end{array} \right] \qquad (A.19)$$

But by using (2.8) individually in each term,

$$\forall z \in E^{(G^{(A_1)})}, def(\varepsilon^{(A_2)}(\varepsilon^{(A_1)}(z))) \Rightarrow$$
$$\left[ \begin{array}{l} \nu^{(A_2)}(ini^{(H^{(A_1)})}(\varepsilon^{(A_1)}(z))) = ini^{(H^{(A_2)})}(\varepsilon^{(A_2)}(\varepsilon^{(A_1)}(z)))\wedge \\ \nu^{(A_2)}(ter^{(H^{(A_1)})}(\varepsilon^{(A_1)}(z))) = ter^{(H^{(A_2)})}(\varepsilon^{(A_2)}(\varepsilon^{(A_1)}(z))) \end{array} \right] \qquad (A.20)$$

which using the fact that $H^{(A_1)} = G^{(A_2)}$, it can be rewritten:

$$\forall z \in E^{(G^{(A_1)})}, def(\varepsilon^{(A_2)}(\varepsilon^{(A_1)}(z))) \Rightarrow$$
$$\left[ \begin{array}{l} \nu^{(A_2)}(ini^{(G^{(A_2)})}(\varepsilon^{(A_1)}(z))) = ini^{(H^{(A_2)})}(\varepsilon^{(A_2)}(\varepsilon^{(A_1)}(z)))\wedge \\ \nu^{(A_2)}(ter^{(G^{(A_2)})}(\varepsilon^{(A_1)}(z))) = ter^{(H^{(A_2)})}(\varepsilon^{(A_2)}(\varepsilon^{(A_1)}(z))) \end{array} \right] \qquad (A.21)$$

and using again (2.8) yields the tautology:

$$\forall z \in E^{(G^{(A_1)})}, def(\varepsilon^{(A_2)}(\varepsilon^{(A_1)}(z))) \Rightarrow$$
$$\left[ \begin{array}{l} ini^{(H^{(A_2)})}(\varepsilon^{(A_2)}(\varepsilon^{(A_1)}(z))) = ini^{(H^{(A_2)})}(\varepsilon^{(A_2)}(\varepsilon^{(A_1)}(z)))\wedge \\ ter^{(H^{(A_2)})}(\varepsilon^{(A_2)}(\varepsilon^{(A_1)}(z))) = ter^{(H^{(A_2)})}(\varepsilon^{(A_2)}(\varepsilon^{(A_1)}(z))) \end{array} \right] \qquad (A.22)$$

∎

## A.2 Composition of Abstractions in *A Graph* is Associative

This section demonstrates what is stated in constraint (2.13):

$$\forall G, H, J, K \in \Theta,$$
$$\forall A_1 = (G, H, \nu_1, \varepsilon_1), A_2(H, J, \nu_2, \varepsilon_2), A_3 = (J, K, \nu_3, \varepsilon_3) \in \nabla, \qquad (A.23)$$
$$(A_3 \Diamond A_2)\Diamond A_1 = A_3 \Diamond (A_2 \Diamond A_1)$$

This equality can be decomposed into abstraction of vertexes and abstraction of edges (since function $\Diamond$ has been defined in that way) as long as the involved abstractions are defined[1]:

---

[1] We denote by $\lambda_0^{(A_i \Diamond A_j)}$ the composition of function $\lambda$ (which can be either $\varepsilon$ or $\nu$) through two abstractions $A_i$ and $A_j$.

$$\forall a \in V^{(G^{(A_1)})}, \left[ \begin{array}{l} def(\nu^{(A_1)}(a)) \wedge def(\nu_{\circ}^{(A_3 \Diamond A_2)}(\nu^{(A_1)}(a))) \wedge \\ def(\nu_{\circ}^{(A_2 \Diamond A_1)}(a)) \wedge def(\nu^{(A_3)}(\nu^{(A_2 \Diamond A_1)}(a))) \end{array} \right] \Rightarrow \\ \left[ \nu^{(A_3 \Diamond A_2)}(\nu^{(A_1)}(a)) = \nu^{(A_3)}(\nu_{\circ}^{(A_2 \Diamond A_1)}(a)) \right] \tag{A.24}$$

$$\forall z \in E^{(G^{(A_1)})}, \left[ \begin{array}{l} def(\varepsilon^{(A_1)}(z)) \wedge def(\varepsilon_{\circ}^{(A_3 \Diamond A_2)}(\varepsilon^{(A_1)}(z))) \wedge \\ def(\varepsilon_{\circ}^{(A_2 \Diamond A_1)}(z)) \wedge def(\varepsilon^{(A_3)}(\varepsilon^{(A_2 \Diamond A_1)}(z))) \end{array} \right] \Rightarrow \\ \left[ \varepsilon^{(A_3 \Diamond A_2)}(\varepsilon^{(A_1)}(z)) = \nu^{(A_3)}(\varepsilon_{\circ}^{(A_2 \Diamond A_1)}(z)) \right] \tag{A.25}$$

Substituting $\nu_{\circ}$ in (A.24) by its definition yields a tautology, as well as substituting $\varepsilon_{\circ}$ in (A.25):

$$\forall a \in V^{(G^{(A_1)})}, \left[ \begin{array}{c} def(\nu^{(A_1)}(a)) \wedge def(\nu^{(A_2)}(\nu^{(A_1)}(a))) \wedge \\ def(\nu^{(A_3)}(\nu^{(A_2)}(\nu^{(A_1)}(a)))) \end{array} \right] \Rightarrow \\ \left[ \nu^{(A_3)}(\nu^{(A_2)}(\nu^{(A_1)}(a))) = \nu^{(A_3)}(\nu^{(A_2)}(\nu^{(A_1)}(a))) \right] \tag{A.26}$$

$$\forall z \in E^{(G^{(A_1)})}, \left[ \begin{array}{c} def(\varepsilon^{(A_1)}(z)) \wedge def(\varepsilon^{(A_2)}(\varepsilon^{(A_1)}(a))) \wedge \\ def(\varepsilon^{(A_3)}(\varepsilon^{(A_2)}(\varepsilon^{(A_1)}(z)))) \end{array} \right] \Rightarrow \\ \left[ \varepsilon^{(A_3)}(\varepsilon^{(A_2)}(\varepsilon^{(A_1)}(z))) = \varepsilon^{(A_3)}(\varepsilon^{(A_2)}(\varepsilon^{(A_1)}(z))) \right] \tag{A.27}$$

■

## A.3 Composition and Identity of Abstractions in *A Graph*

This section demonstrates what is stated in constraint (2.14):

$$\forall G, H \in \Theta, \forall A = (G, H, \nu, \varepsilon) \\ A \Diamond i(G) = A = i(G) \Diamond A \tag{A.28}$$

Substituting by the definition of each function and separating abstraction of vertexes from abstraction of edges:

$$\forall a \in V^{(G)}, \left[ def(\nu^{(A)}(a)) \wedge def(\nu_{\circ}^{(A \Diamond i(G))}(a)) \wedge def(\nu_{\circ}^{(i(G) \Diamond A)}(a)) \right] \Rightarrow \\ \left[ \nu_{\circ}^{(A \Diamond i(G))}(a) = \nu^{(A)}(a) = \nu_{\circ}^{(i(G) \Diamond A)}(a) \right] \tag{A.29}$$

$$\forall z \in E^{(G)}, \left[ def(\varepsilon^{(A)}(z)) \wedge def(\varepsilon_{\circ}^{(A\Diamond i(G))}(z)) \wedge def(\varepsilon_{\circ}^{(i(G)\Diamond A)}(z)) \right] \Rightarrow$$
$$\left[ \varepsilon_{\circ}^{(A\Diamond i(G))}(z) = \varepsilon^{(A)}(z) = \nu_{\circ}^{(i(G)\Diamond A)}(z) \right] \qquad (A.30)$$

Firstly, (A.29) can be deployed using the definition of $\nu_{\circ}$:

$$\forall a \in V^{(G)}, \left[ def(\nu^{(A)}(a)) \wedge def(\nu(A)(\nu^{(i(G))}(a))) \wedge def(\nu^{(i(G))}(\nu^{(A)}(a))) \right] \Rightarrow$$
$$\left[ \nu^{(A)}(\nu^{(i(G))}(a)) = \nu^{(A)}(a) = \nu^{(i(G))}(\nu^{(A)}(a)) \right]$$
$$(A.31)$$

But by definition (2.11), $\nu^{(i(G))}(x) = \nu_G(x) = x, \forall x$, so the equality is satisfied.

Deploying $\varepsilon_{\circ}$ in (A.30) and following (2.11) in the same manner, the equality is also satisfied.

$\blacksquare$

## A.4 Composition of Refinements in *RGraph* is a Refinement

This section demonstrates what is stated in (2.23), that is that $R_2 \bullet R_1$ satisfies constraints for refinements of graphs. The constraints that function $\bullet$ must satisfy for being considered a refinement are (2.15-2.18), that we repeat here for convenience:

$$\mu : V^{(G)} \to 2^{V^{(H)}} \ is \ a \ total \ function. \qquad (A.32)$$

$$\alpha : E^{(G)} \to 2E^{(H)} \ is \ a \ total \ function. \qquad (A.33)$$

$$\forall a \neq b \in V^{(G)}, \ \mu(a) \cap \mu(b) = \phi$$
$$\forall y \neq z \in E^{(G)}, \ \alpha(y) \cap \alpha(z) = \phi \qquad (A.34)$$

$$\forall z \in E^{(G)}, \forall y \in \alpha(z), \left[ \begin{array}{l} ini^{(H)}(y) \in \mu(ini^{(G)}(z)) \wedge \\ ter^{(H)}(y) \in \mu(ter^{(G)}(z)) \end{array} \right] \qquad (A.35)$$

*Demonstration of (A.32) and (A.33)*

These two constraints are trivially satisfied by the composition function: since $\mu^{(R_1)}$ is total, $\mu^{(R_2)} \times \mu^{(R_1)}$ must be too because definition of $\times$ in (2.23) covers the complete dominion of $\mu^{(R_1)}$. The same is valid for $\alpha$.

$\blacksquare$

*Demonstration of (A.34)*

That constraint is, for $R_2 \bullet R_1$:

$$\forall a \neq b \in V^{(G^{(R_1)})}, \left[\mu^{(R_2)} \times \mu^{(R_1)}(a)\right] \cap \left[\mu^{(R_2)} \times \mu^{(R_1)}(b)\right] = \phi$$
$$\forall y \neq z \in E^{(G^{(R_1)})}, \left[\alpha^{(R_2)} \times \alpha^{(R_1)}(y)\right] \cap \left[\alpha^{(R_2)} \times \alpha^{(R_1)}(z)\right] = \phi \tag{A.36}$$

If we substitute the operators $\times$ by their definitions in (2.23):

$$\forall a \neq b \in V^{(G^{(R_1)})}, \left[\bigcup^{\forall c \in \mu^{(R_1)}(a)} \mu^{(R_2)}(c)\right] \cap \left[\bigcup^{\forall d \in \mu^{(R_1)}(b)} \mu^{(R_2)}(d)\right] = \phi$$

$$\forall y \neq z \in E^{(G^{(R_1)})}, \left[\bigcup^{\forall x \in \alpha^{(R_1)}(y)} \alpha^{(R_2)}(x)\right] \cap \left[\bigcup^{\forall w \in \alpha^{(R_1)}(z)} \mu^{(R_2)}(w)\right] = \phi \tag{A.37}$$

But by set algebra [126], we know that the following is true for any sets $A_i, B_i$:

$$\left(\cup^{i \in 1...n} A_i\right) \cap \left(\cup^{j \in 1...m} B_j\right) = \cup^{i \in 1...n, j \in 1...m}(A_i \cap B_j) \tag{A.38}$$

So we can reorder (A.37):

$$\forall a \neq b \in V^{(G^{(R_1)})}, \bigcup^{\forall c \in \mu^{(R_1)}(a), \forall d \in \mu^{(R_1)}(b)} (\mu^{(R_2)}(c) \cap \mu^{(R_2)}(d)) = \phi$$

$$\forall y \neq z \in E^{(G^{(R_1)})}, \bigcup^{\forall x \in \alpha^{(R_1)}(y), \forall w \in \alpha^{(R_1)}(z)} (\alpha^{(R_2)}(x) \cap \alpha^{(R_2)}(w)) = \phi \tag{A.39}$$

But by definition of $\mu$ and $\alpha$ we know that:

$$\forall a \neq b \in V^{(G^{(R_1)})}, \mu^{(R_1)}(a) \cap \mu^{(R_1)}(b) = \phi$$
$$\forall c \neq d \in V^{(G^{(R_2)})}, \mu^{(R_2)}(c) \cap \mu^{(R_2)}(d) = \phi$$
$$\forall y \neq z \in E^{(G^{(R_1)})}, \alpha^{(R_1)}(y) \cap \alpha^{(R_1)}(z) = \phi$$
$$\forall x \neq w \in E^{(G^{(R_2)})}, \alpha^{(R_2)}(x) \cap \alpha^{(R_2)}(w) = \phi \tag{A.40}$$

Therefore, (A.39) is trivially true.

∎

*Demonstration of (A.35)*

For $R_2 \bullet R_1$, that constraint is:

$$\forall z \in E^{(G^{(R_1)})}, \forall y \in \alpha^{(R_2)} \times \alpha^{(R_1)}(z),$$
$$\left[ \begin{array}{l} ini^{(H^{(R_2)})}(y) \in \mu^{(R_2)} \times \mu^{(R_1)}(ini^{(G^{(R_1)})}(z)) \wedge \\ ter^{(H^{(R_2)})}(y) \in \mu^{(R_2)} \times \mu^{(R_1)}(ter^{(G^{(R_1)})}(z)) \end{array} \right] \tag{A.41}$$

which can be rewritten using the definitions of operators $\times$:

$$\forall z \in E^{(G^{(R_1)})}, \forall y \in \bigcup_{}^{\forall x \in \alpha^{(R_1)}(z)} \alpha^{(R_2)}(x),$$
$$\left[ \begin{array}{l} ini^{(H^{(R_2)})}(y) \in \bigcup_{}^{\forall b \in \mu^{(R_1)}(ini^{(G^{(R_1)})}(z))} \mu^{(R_2)}(b) \wedge \\ ter^{(H^{(R_2)})}(y) \in \bigcup_{}^{\forall c \in \mu^{(R_1)}(ter^{(G^{(R_1)})}(z))} \mu^{(R_2)}(c) \end{array} \right] \tag{A.42}$$

We know that (2.18) holds for $R_1$ and $R_2$ separately:

$$\forall z \in E^{(G^{(R_1)})}, \forall y \in \alpha^{(R_1)}(z), \left[ \begin{array}{l} ini^{(H^{(R_1)})}(y) \in \mu^{(R_1)}(ini^{(G^{(R_1)})}(z)) \wedge \\ ter^{(H^{(R_1)})}(y) \in \mu^{(R_1)}(ter^{(G^{(R_1)})}(z)) \end{array} \right]$$
$$\forall x \in E^{(G^{(R_2)})}, \forall w \in \alpha^{(R_2)}(x), \left[ \begin{array}{l} ini^{(H^{(R_2)})}(w) \in \mu^{(R_2)}(ini^{(G^{(R_2)})}(x)) \wedge \\ ter^{(H^{(R_2)})}(w) \in \mu^{(R_2)}(ter^{(G^{(R_2)})}(x)) \end{array} \right] \tag{A.43}$$

In particular, if the second equation of (A.43) holds $\forall x \in E^{(G^{(R_2)})}$, it must hold $\forall x \in \alpha^{(R_1)}(z)$, given some $z \in E^{(G^{(R_1)})}$:

$$\forall z \in E^{(G^{(R_1)})}, \forall x \in \alpha^{(R_1)}(z), \forall w \in \alpha^{(R_2)}(x),$$
$$\left[ \begin{array}{l} ini^{(H^{(R_2)})}(w) \in \mu^{(R_2)}(ini^{(G^{(R_2)})}(x)) \wedge \\ ter^{(H^{(R_2)})}(w) \in \mu^{(R_2)}(ter^{(G^{(R_2)})}(x)) \end{array} \right] \tag{A.44}$$

Now notice that by the first equation of (A.43),

$$\forall x \in \alpha^{(R_1)}(z),$$
$$\left[ ini^{(G^{(R_2)})}(x) \in \mu^{(R_1)}(z) \right] \wedge \left[ ter^{(G^{(R_2)})}(x) \in \mu^{(R_1)}(ter^{(G^{(R_1)})}(z)) \right] \tag{A.45}$$

Applying $\mu^{(R_2)}$ to (A.45):

$$\forall x \in \alpha^{(R_1)}(z),$$

$$\left[ \mu^{(R_2)}(ini^{(G^{(R_2)})}(x) \subseteq \bigcup^{\forall b \in \mu^{(R_1)}(ini^{(G^{(R_1)})}(z))} \mu^{(R_2)}(b) \right] \wedge$$
$$\left[ \mu^{(R_2)}(ter^{(G^{(R_2)})}(x) \subseteq \bigcup^{\forall c \in \mu^{(R_1)}(ter^{(G^{(R_1)})}(z))} \mu^{(R_2)}(c) \right] \tag{A.46}$$

Using (A.46) we can rewrite (A.44):

$$\forall z \in E^{(G^{(R_1)})}, \forall x \in \alpha^{(R_1)}(z), \forall w \in \alpha^{(R_2)}(x),$$

$$\left[ \begin{array}{l} ini^{(H^{(R_2)})}(w) \in \bigcup^{\forall b \in \mu^{(R_1)}(ini^{(G^{(R_1)})}(z))} \mu^{(R_2)}(b) \wedge \\ ter^{(H^{(R_2)})}(w) \in \bigcup^{\forall c \in \mu^{(R_1)}(ter^{(G^{(R_1)})}(z))} \mu^{(R_2)}(c) \end{array} \right] \tag{A.47}$$

Since (A.47) does not use $x$ anymore, it can be rewritten:

$$\forall z \in E^{(G^{(R_1)})}, \forall w \in \bigcup^{\forall x \in \mu^{(R_1)}(z)} \alpha^{(R_2)}(x),$$

$$\left[ \begin{array}{l} ini^{(H^{(R_2)})}(w) \in \bigcup^{\forall b \in \mu^{(R_1)}(ini^{(G^{(R_1)})}(z))} \mu^{(R_2)}(b) \wedge \\ ter^{(H^{(R_2)})}(w) \in \bigcup^{\forall c \in \mu^{(R_1)}(ter^{(G^{(R_1)})}(z))} \mu^{(R_2)}(c) \end{array} \right] \tag{A.48}$$

which is (A.42), what we wanted to demonstrate.

$\blacksquare$

## A.5 Composition of Refinements in *RGraph* is Associative

This section demonstrates what is stated in (2.24):

$$\forall G, H, J, K \in \Theta,$$

$$\left[ \begin{array}{c} \forall R_1 = (G, H, \mu_1, \alpha_1), R_2 = (H, J, \mu_2, \alpha_2), \\ R_3 = (J, K, \mu_3, \alpha_3) \in \Delta, \\ (R_3 \bullet R_2) \bullet R_1 = R_3 \bullet (R_2 \bullet R_1) \end{array} \right]$$

The equality can be decomposed into refinement of vertexes and refinement of edges (since function $\bullet$) has been defined in that way):

$$\forall a \in V^{(G^{(R_1)})}, \left(\left(\mu^{(R3)} \times \mu^{(R2)}\right) \times \mu^{(R1)}\right)(a) = \left(\mu^{(R3)} \times \left(\mu^{(R2)} \times \mu^{(R1)}\right)\right)(a)$$
$$\forall z \in E^{(G^{(R_1)})}, \left(\left(\alpha^{(R3)} \times \alpha^{(R2)}\right) \times \alpha^{(R1)}\right)(z) = \left(\alpha^{(R3)} \times \left(\alpha^{(R2)} \times \alpha^{(R1)}\right)\right)(z)$$
$$\text{(A.49)}$$

Substituting $\times$ in (A.49) by its definition yields the following tautologies:

$$\forall a \in V^{(G^{(R_1)})}, \bigcup_{\forall b \in \mu^{(R_1)}(a)} \mu^{(R_3)} \times \mu^{(R_2)}(b) =$$
$$\mu^{(R_3)} \times \left(\bigcup_{\forall c \in \mu^{(R_1)}(a)} \mu^{(R_2)}(c)\right) = \bigcup_{\forall c \in \mu^{(R_1)}(a)} \mu^{(R_3)} \times \mu^{(R_2)}(c)$$
$$\text{(A.50)}$$
$$\forall z \in E^{(G^{(R_1)})}, \bigcup_{\forall y \in \alpha^{(R_1)}(z)} \alpha^{(R_3)} \times \alpha^{(R_2)}(y) =$$
$$\alpha^{(R_3)} \times \left(\bigcup_{\forall x \in \alpha^{(R_1)}(z)} \alpha^{(R_2)}(x)\right) = \bigcup_{\forall x \in \alpha^{(R_1)}(z)} \alpha^{(R_3)} \times \alpha^{(R_2)}(x)$$

$\blacksquare$

## A.6 Composition and Identity of Refinements in *RGraph*

This section demonstrates what is stated in (2.25):
$$\forall G, H \in \Theta, \forall R = (G, H, \mu, \alpha),$$

$$R \bullet i(G) = R = i(G) \bullet R$$

Separating refinement of vertexes from refinement of edges:

$$\forall a \in V^{(G)}, \mu^{(R)} \times \mu^{(I(G))}(a) = \mu^{(R)}(a) = \mu^{(I(G))} \times \mu^{(R)}(a)$$
$$\forall z \in E^{(G)}, \alpha^{(R)} \times \alpha^{(I(G))}(z) = \alpha^{(R)}(z) = \alpha^{(I(G))} \times \alpha^{(R)}(z) \qquad \text{(A.51)}$$

This can be deployed using the definition of $\times$:

$$\forall a \in V^{(G)}, \bigcup_{\forall b \in \mu^{(I(G))}(a)} \mu^{(R)}(b) = \mu^{(R)}(a) = \bigcup_{\forall c \in \mu^{(R)}(a)} \mu^{(I(G))}(c)$$
$$\forall z \in E^{(G)}, \bigcup_{\forall y \in \alpha^{(I(G))}(z)} \alpha^{(R)}(y) = \alpha^{(R)}(z) = \bigcup_{\forall w \in \alpha^{(R)}(z)} \alpha^{(I(G))}(z) \qquad \text{(A.52)}$$

Which can be deployed again using the definition for i($G$):

$$\forall a \in V^{(G)}, \quad \bigcup_{\forall b \in \mu_G(a)} \mu^{(R)}(b) = \mu^{(R)}(a) = \bigcup_{\forall c \in \mu^{(R)}(a)} \mu_G(c)$$
$$\forall z \in E^{(G)}, \quad \bigcup_{\forall y \in \alpha_G(z)} \alpha^{(R)}(y) = \alpha^{(R)}(z) = \bigcup_{\forall w \in \alpha^{(R)}(z)} \alpha_G(w) \tag{A.53}$$

And:

$$\forall a \in V^{(G)}, \quad \bigcup_{\forall b \in \{a\} \mu^{(R)}(b) = \mu^{(R)}(a) =} \bigcup_{\forall c \in \mu^{(R)}} (a) \quad \{c\}$$
$$\forall z \in E^{(G)}, \quad \bigcup_{\forall y \in \{z\} \alpha^{(R)}(y) = \alpha^{(R)}(z) =} \bigcup_{\forall w \in \alpha^{(R)}} (z) \quad \{w\} \tag{A.54}$$

Which is a tautology:

$$\forall a \in V^{(G)}, \mu^{(R)}(a) = \mu^{(R)}(a)$$
$$\forall z \in E^{(G)}, \alpha^{(R)}(z) = \alpha^{(R)}(z) \tag{A.55}$$

∎

# B

# Demonstration that $\Psi$ is a Functor from *CAVGraph** to *AState*

## B.1 $\Psi$ Preserves Identity

In this section we demonstrate that given a graph $G \in \Theta^*$:

$$\Psi(\mathrm{i}(G)) = \mathrm{i}(\Psi(G))$$

for the functor $\Psi$ between *CAVGraph** and *AState*.

On the one hand, by equation 2.11: $\Psi(\mathrm{i}(G)) = \Psi_a(A_1)$, where $A_1 = (G, G, \nu_G, \varepsilon_G)$:

$$\nu_G : V^{(G)} \to V^{(G)} \qquad \varepsilon_G : E^{(G)} \to E^{(G)}$$
$$\forall a \in V^{(G)}, \nu_G(a) = a \ \forall z \in E^{(G)}, \varepsilon_G(z) = z$$

On the other hand, applying the identity function of *AState* (equation 3.5) to $\Psi_o(G)$:

$$\mathrm{i}(\Psi_o(G)) = A_{s1} = (\Psi_o(G), \Psi_o(G), \xi, \pi)$$

where

$$\xi : SN(\Psi_o(G)) \to SN(\Psi_o(G)) \ \pi : SP(\Psi_o(G)) \to SP(\Psi_o(G))$$
$$\forall p \in SN(\Psi_o(G)), \xi(p) = p \qquad \forall a \in SP(\Psi_o(G)), \pi(a) = a$$

Thus, we must demonstrate that

$$\Psi_a(A_1) = A_{s1}$$

By definition 3.14:

$$\Psi_a(A_1) = \Psi_a((G, G, \nu_G, \varepsilon_G)) = (\Psi_o(G), \Psi_o(G), \tau, \kappa), \ where$$

$$\tau : \Upsilon \to \Upsilon$$
$$\forall z \in E^{(G)} : def(\varepsilon_G(z)), \tau(\Gamma_{e1}(z)) = \Gamma_{e1}(\varepsilon_G(z)) = \Gamma_{e1}(z)$$

$$\kappa : \Upsilon \to \Upsilon$$
$$\forall a \in V^{(G)} : def(\nu_G(a), \kappa(\Gamma_v(a)) = \Gamma_v(\nu_G(a)) = \Gamma_v(a)$$

That is, $\tau = \xi^{(A_{s1})}$ and $\kappa = \pi(A_{s1})$, so the identity is preserved. ∎

## B.2 $\Psi$ Preserves Composition of Arrows

This section demonstrates that $\Psi$ preserves the composition of arrows, that is:

$$\forall A_1, A_2 \in \triangledown : def(A_2 \lozenge A_1), \Psi(A_2 \lozenge A_1) = \Psi(A_2) \lozenge \Psi(A_1)$$

Let be $A_1 = (G, H, \nu_1, \varepsilon_1)$ and $A_2(H, J, \nu_2, \varepsilon_2)$ two abstractions of *CAVGraph**. By definition 2.12, $A_2 \lozenge A_1 = (G, J, \nu_o, \varepsilon_o)$, where the two abstraction functions are defined by mathematical composition of functions: $\nu_o = \nu_2 \circ \nu_1$ and $\varepsilon_o = \varepsilon_2 \circ \varepsilon_1$.

Applying the functor $\Psi_a$ to the composition $A_2 \lozenge A_1$, we obtain:

$$\Psi_a(A_2 \lozenge A_1) = \Psi_a((G, J, \nu_o, \varepsilon_o)) = (\Psi_o(G), \Psi_o(J), \tau_o, \kappa_o)$$

where functions $\tau_o$ and $\kappa_o$ are defined in terms of $\varepsilon_o$ and $\nu_o$ respectively:

$$\tau_o : \Upsilon \to \Upsilon$$
$$\forall z \in E^{(G)} : def(\varepsilon_o(z)), \tau_o(\Gamma_{e1}(z)) = \Gamma_{e1}(\varepsilon_o(z)) = \Gamma_{e1}(\varepsilon_2(\varepsilon_1(z)))$$

$$\kappa_o : \Upsilon \to \Upsilon$$
$$\forall a \in V^{(G))} : def(\nu_o(a), \kappa_o(\Gamma_v(a)) = \Gamma_v(\nu_o(a)) = \Gamma_v(\nu_2(\nu_1((a))))$$

On the other hand, we must compute $\Psi_a(A_2) \lozenge \Psi_a(A_1)$:

$$\Psi_a(A_2) = (\Psi_o(H), \Psi_o(J), \tau_2, \kappa_2)$$
$$\forall z \in E^{(H)} : def(\varepsilon_2(z)), \tau_2(\Gamma_{e1}(z)) = \Gamma_{e1}(\varepsilon_2(z))$$
$$\forall a \in V^{(H)} : def(\nu_2(a), \kappa_2(\Gamma_v(a)) = \Gamma_v(\nu_2(a))$$

$$\Psi_a(A_1) = (\Psi_o(G), \Psi_o(H), \tau_1, \kappa_1)$$
$$\forall z \in E^{(G)} : def(\varepsilon_1(z)), \tau_1(\Gamma_{e1}(z)) = \Gamma_{e1}(\varepsilon_1(z))$$
$$\forall a \in V^{(G)} : def(\nu_1(a), \kappa_1(\Gamma_v(a)) = \Gamma_v(\nu_1(a))$$

$$\Psi_a(A_2) \lozenge \Psi_a(A_1) = (\Psi_o(G), \Psi_o(J), \xi_3, \pi_3)$$

by definition 3.6 of composition of abstractions in *AState*. It remains to demonstrate that functions $\xi_3$ and $\pi_3$ are equivalent to $\tau_o$ and $\kappa_o$ respectively. But, by definition 3.6:

$$\xi_3 = \tau_2 \circ \tau_1 :$$
$$\left[\forall z \in E^{(G)} : def(\varepsilon_1(z)), \varepsilon_1(z) \in E^{(H)}, \tau_1(\Gamma_{e1}(z)) = \Gamma_{e1}(\varepsilon_1(z))\right] \wedge$$
$$\left[\forall z \in E^{(H)} : def(\varepsilon_2(z)), \tau_2(\Gamma_{e1}(z)) = \Gamma_{e1}(\varepsilon_2(z))\right] \Rightarrow$$
$$\forall z \in E^{(G)} : def(\varepsilon_1(z)) \wedge def(\varepsilon_2(z)) \wedge def(\varepsilon_2(\varepsilon_1(z)))$$
$$\xi_3(\Gamma_{e1}(z)) = \tau_2(\tau_1(\Gamma_{e1}(z))) = \tau_2(\Gamma_{e1}(\varepsilon_1(z))) = \Gamma_{e1}(\varepsilon_2(\varepsilon_1(z)))$$

$$\pi_3 = \kappa_2 \circ \kappa_1 :$$
$$\left[\forall a \in V^{(G)} : def(\nu_1(a)), \nu_1(a) \in V^{(G)}, \kappa_1(\Gamma_v(a)) = \Gamma_v(\nu_1(a))\right] \wedge$$
$$\left[\forall a \in V^{(H)} : def(\nu_2(a)), \kappa_2(\Gamma_v(a)) = \Gamma_v(\nu_2(a))\right] \Rightarrow$$
$$\forall a \in V^{(G)} : def(\nu_1(a)) \wedge def(\nu_2(a)), \wedge def(\nu_2(\nu_1(a)))$$
$$\pi_3(\Gamma_v(a)) = \kappa_2(\kappa_1(\Gamma_v(a))) = \kappa_2(\Gamma_v(\nu_1(a))) = \Gamma_v(\nu_2(\nu_1(a))) \qquad \blacksquare$$

# C

# Planning Domain

```
(define (domain experiment)
  (:requirements :typing :fluents)
  (:types          LOCATION OBJECT)
  (:constants      Robot - OBJECT)

  (:predicates
        (at ?obj - OBJECT ?loc - LOCATION)
        (free ?obj - OBJECT)
        (on ?sup ?inf - OBJECT)
        (in ?obj1 ?obj2 - OBJECT)
        (free-robot)
        (taken ?obj - OBJECT)
        (closed ?obj - OBJECT)
        (opened ?obj - OBJECT)
        (Nav ?x ?y - LOCATION)
        (charge-station ?x -LOCATION)
        (separated ?x ?y - LOCATION)
        (connected ?x ?y - LOCATION ?z - OBJECT)
    )

    (:functions
        (time-to-move ?l1 ?l2 - location)
        (battery-to-move ?l1 ?l2 - location)
        (battery)
        (time)
    )

(:action GO
  :parameters
    (
```

```
     ?loc-from - LOCATION
     ?loc-to - LOCATION
     )
  :precondition
   (and      (at Robot ?loc-from)
    (Nav ?loc-from ?loc-to)
    (and (not (separated ?loc-from ?loc-to))
    (not (separated ?loc-to ?loc-from)))
    (> (battery) (battery-to-move ?loc-from ?loc-to))
    )
  :effect
   (and (not (at Robot ?loc-from)) (at Robot ?loc-to)
    (increase (time) (time-to-move ?loc-from ?loc-to))
    (decrease (battery) (battery-to-move ?loc-from ?loc-to))
    )
)

(:action RECHARGE-LOW
 :parameters
 (
  ?loc - LOCATION
  )
  :precondition
  (and (charge-station ?loc)
       (at Robot ?loc)
  )
  :effect
  (and
    (increase (battery) 5)
    (increase (time) 5)
    )
)

(:action RECHARGE-MED
 :parameters
 (
  ?loc - LOCATION
  )
  :precondition
  (and (charge-station ?loc)
       (at Robot ?loc)
  )
  :effect
  (and
    (increase (battery) 10)
```

```
      (increase (time) 10)
      )
  )

(:action RECHARGE-HIGH
 :parameters
 (
  ?loc - LOCATION
  )
  :precondition
  (and (charge-station ?loc)
       (at Robot ?loc)
  )
  :effect
 (and
    (increase (time) 15)
    (increase (battery) 15)
  )
 )

(:action TAKEOUT
  :parameters
   (
     ?obj - OBJECT
    ?obj2 - OBJECT
     ?loc - LOCATION
     )
  :precondition
    ( and  (at Robot ?loc)
           (at ?obj2 ?loc)
          (in ?obj ?obj2)
        (opened ?obj2)
          (free-robot)
     )
  :effect
    (and (not (free-robot)) (not (at ?obj ?loc)) (taken ?obj)
      (not (in ?obj ?obj2)))
 )

(:action TAKE
  :parameters
   (
     ?obj - OBJECT
     ?loc - LOCATION
     )
```

```
    :precondition
      ( and  (at Robot ?loc)
             (at ?obj ?loc)
          (free ?obj)
          (free-robot)
       )
    :effect
      (and (not (free ?obj)) (not (free-robot)) (not (at ?obj
            ?loc)) (taken ?obj))
 )

(:action REMOVE
   :parameters
    (
      ?obj - OBJECT
    ?obj2 - OBJECT
      ?loc - LOCATION
      )
   :precondition
     ( and (at Robot ?loc)
            (at ?obj ?loc)
         (on ?obj ?obj2)
    (free ?obj)
    (free-robot)
      )
   :effect
     (and (not (free ?obj)) (not (free-robot)) (not (at ?obj
           ?loc)) (taken ?obj))
 )

(:action GIVE
   :parameters
    (
     ?obj - OBJECT
     ?loc - LOCATION
    )
   :precondition
   ( and    (at Robot ?loc)
    (taken ?obj)
   )
   :effect
   ( and    (not (taken ?obj)) (at ?obj ?loc) (free-robot) )
 )

(:action PUTDOWN
```

```
:parameters
  (
    ?objsup - OBJECT
    ?objinf  - OBJECT
    ?loc - LOCATION
  )
:precondition
  (and    (at Robot ?loc) (at ?objsup ?loc) (at ?objinf
           ?loc) (on ?objsup ?objinf) (free ?objsup))
:effect
  (and (not (on ?objsup ?objinf)) (free ?objinf))
)
)
```

# References


1. A. Doucet A., S. Godsill, and C. Andrieu, *On sequential monte carlo sampling methods for bayesian filtering*, Statistics and Computing **10** (2000), 197–208.
2. A.V. Aho, J.E. Hopcroft, and J.D. Ullman, *Data structures and algorithms*, Addison-Wesley Publishing Co., Massachusetts, 1983.
3. P. Aigner and B. McCarragher, *Human integration into robot control utilizing potential fields*, IEEE Int. Conf. on Robotics and Automation, Albuquerque, NM, 1997, pp. 291–296.
4. R. Alami, R. Chatila, S. Fleury, M. Ghallab, and F. Ingrand, *An architecture for autonomy*, International Journal of Robotics Research **17** (1998), no. 4, 315–337.
5. J. G. Allen, R. Xu, and J. Jin, *Object tracking using camshift algorithm and multiple quantized feature spaces*, Proc. of the Pan-Sydney area workshop on Visual information processing, Australia, 2004.
6. R.C. Arkin, *Cooperation without communication: Multi-agent schema based robot navigation*, Journal of Robotic Systems **9** (1992), no. 3, 351–364.
7. R.C. Arkin, E.M. Riseman, and A. Hansen, *Aura: An architecture for vision-based robot navigation*, DARPA Image Understanding Workshop, Los Angeles, CA, 1987, pp. 417–431.
8. M. Asada, *Map building for a mobile robot from sensory data*, IEEE Trans. on Systems, Man, and Cybernetics **37** (1990), no. 6, 1326–1336.
9. T. Bäck, U. Hammel, and H. Schwefel, *Evolutionary computation: Comments on the history and current state*, IEEE Trans. on Evolutionary Computation **1** (1997), no. 1, 3–17.
10. M. Baioletti, S. Marcugini, and A. Milani, *DPPlan: An algorithm for fast solutions extraction from a planning graph*, 5th International Conference on Artificial Planning Systems, 2000, pp. 13–21.
11. K. Baker, *Introduction to sequencing and scheduling*, Wiley, 1974.
12. M.H. Bickhard and L. Terveen, *Foundational issues in artificial intelligence and cognitive science*, Elsevier Science, 1995.
13. J.L. Blanco, A. Cruz, Galindo, Fernández-Madrigal, and J. González, *Towards a multi-agent software architecture for human-robot integration*, 4th International Workshop on Practical Applications of Agents and Multiagent Systems (IWPAAMS), León (Spain), 2005, pp. 235–244.

14. J.L. Blanco, J. González, and J.A. Fernández-Madrigal, *Consistent observation grouping for generating metric-topological maps that improves robot localization*, IEEE Int. Conf. on Robotics and Automation, Orlando, May 2006.

15. ______, *A consensus-based approach for estimating the observation likelihood of accurate range sensors*, IEEE Int. Conf. on Robotics and Automation, Rome, Italy, Apr 2007.

16. A.L. Blum and M.L. Furst, *Fast planning through planning graph analysis*, Artificial Intelligence **90** (1997), 281–300.

17. A. Bonarini, M. Matteucci, and M. Restelli, *Concepts for anchoring in robotics. AI*IA: Advances in artificial intelligence*, Springer-Verlag, 2001.

18. M. Bosse, P. M. Newman, J.J. Leonard, and S. Teller, *Slam in large-scale cyclic environments using the atlas framework*, Int. Journal on Robotics Research **23** (2004), no. 12, 1113–1139.

19. M. Bosse, P.M. Newman, J.J. Leonard, M. Soika, W. Feiten, and S.J. Teller, *An atlas framework for scalable mapping*, IEEE Int. Conf. on Robotics and Automation, Taipei (Taiwan), 2003, pp. 1899–1906.

20. S. Brindgsjord, *What robots can and can't be*, Kluwer Academic Publishers, 1992.

21. R.A. Brooks, *A robust layered control system for a mobile robot*, IEEE Journal on Robotics and Automation **2** (1986), no. 1, 14–23.

22. ______, *Architectures for intelligence*, ch. How to build complete creatures rather than isolated cognitive simulators, pp. 225–239, Lawrence Erlbaum Assosiates, Hillsdale, NJ, 1991.

23. M. Broxvall, S. Coradeschi, L. Karlsson, and A. Saffiotti, *Recovery planning for ambiguous cases in perceptual anchoring*, Proc. of the 20th AAAI Conf. Pittsburgh, PA, 2005, pp. 1254–1260.

24. P. Buschka and A. Saffiotti, *A virtual sensor for room detection*, IEEE/RSJ Int. Conf. on Intelligent Robots and Systems, Lausanne, CH, 2002, pp. 637–642.

25. K. Chong and L. Kleeman, *Large scale sonarray mapping using multiple connected local maps*, Proc. Int. Conf. Field Service Robot, 1997, pp. 538–545.

26. H. Choset and K. Nagatani, *Topological simultaneous localization and mapping (SLAM): Toward exact localization without explicit localization*, IEEE Trans. on Robotics and Automation **17** (2001), no. 2, 125–137.

27. J. Christensen, *A hierarchical planner that generates its own hierarchies*, Proc. of the Eighth National Conference on Artificial Intelligence, 1990, pp. 1004–1009.

28. S. Coradeshi and A. Saffiotti, *An introduction to the anchoring problem*, Robotics and Autonomous System **43** (2003), no. 2-3, 85–96.

29. A. Corradini, U. Montanari, F. Rossi, H. Ehrig, R. Heckel, and M. Loewe, *Algebraic approaches to graph transformation: Basic concepts and double pushout approach, Univ. of Pisa*, Tech. report, 1996.

30. J.W. Crandall and M.A. Goodrich, *Experiments in adjustable autonomy*, IEEE Int. Conf. on Systems, Man, and Cybernetcis, Tucson, AZ, 2001, pp. 1624–1629.

31. R.I. Damper, R.L.B. French, and T.W. Scutt, *Arbib: An autonomous robot based on inspirations from biology*, Robotics and Autonomous Systems **31** (2000), 247–274.

32. T. W. Deacon, *The symbolic species: The co-evolution of language and the brain*, W.W. Norton, 1997.

33. A. Delchambre and P. Gaspart, *Kbap: An industrial prototype of knowledge-based assembly planner*, IEEE Intl. Conf. On Robotics and Automation, Nice, France, 1992.

34. J.S. DeLoache, *Becoming symbol-minded*, TRENDS in Cognitive Sciences **8** (2004), no. 2, 66–70.

35. J.S. DeLoache, D.H. Uttal, and K.S. Rosengren, *Scale errors offer evidence for a perception-action dissociation early in life*, Science **304** (2004), 1027–1029.

36. P. Doherty, *The witas integrated software system architecture*, Linkoping Electronic Articles in Computer and Information Science **4** (1999), no. 17.

37. G. Dorais, R.P. Bonasso, D. Kortenkamp, B. Pell, and D. Schreckenghost, *Adjustable autonomy for human-centered autonomous systems on mars*, Proc. of the First International Conference of the Mars Society, 1998, pp. 397–420.

38. S. Eilenberg and S. MacLane, *A general theory of natural equivalences*, Trans. of the American Mathematical Society **58** (1945), 231–294.

39. S.P. Engelson and D.V. McDermott, *Error correction in mobile robot map learning*, IEEE Int. Conf. on Robotics and Automation, Ottawa (Canada), 1992, pp. 2555–2560.

40. K. Erol, J. Hendler, and D. Nau, *Semantics for hierarchical task network planning*, Tech. report, CS-TR-3239, UMIACS-TR-94-31, Computer Science Dept and University of Maryland, 1994.

41. C. Estrada, J. Neira, and J.D. Tardos, *Hierarchical slam: Real-time accurate mapping of large environments*, IEEE Trans. on Robotics **21** (2005), 588–596.

42. E. Fabrizi, G. Oriolo, and G. Ulivi, *Fuzzy logic techniques for autonomous vehicle navigation*, ch. Accurate map building via fusion of laser and ultrasonic range measures, pp. 257–280, Physica / Springer-Verlag, 2001.

43. E. Fabrizi and A. Saffiotti, *Extracting topology-based maps from gridmaps*, IEEE Int. Conf. on Robotics and Automation, San Francisco, CA, 2002, pp. 2972–2978.

44. R. Falcone and C. Castelfranchi, *The human in the loop of a delegated agent: the theory of adjustable social autonomy*, IEEE Transactions on Systems, Man, and Cybernetics, Part A **31** (2001), no. 5, 406–418.

45. J.A. Fernandez, J. Gonzalez, L. Mandow, and J.L. Perez de-la Cruz, *Mobile robot path planning: A multicriteria approach*, Engineering Applications of Artificial Intelligence **12** (1991), no. 4, 543–554.

46. J.L. Fernandez and R.G. Simmons, *Robust execution monitoring for navigation plans*, IEEE/RSJ Int. Conf. on Intelligent Robots and Systems, Victoria B.C (Canada), 1998, pp. 551 – 557.

47. J.A. Fernández-Madrigal, C. Galindo, and J. González, *Assistive navigation of a robotic wheelchair using a multihierarchical model of the environment*, Integrated Computer-Aided Eng. **11** (2004), 309–322.

48. J.A. Fernández-Madrigal, C. Galindo, J. González, E. Cruz, and A. Cruz, *A software engineering approach for the development of heterogeneous robotic applications*, Robotics and Computer-Integrated Manufacturing, (2007 (to appear)).

49. J.A. Fernández-Madrigal and J. González, *Multi-hierarchical representation of large-scale space*, Int. Series on Microprocessor-based and Intell. Systems Eng., vol 24, Kluwer Academic Publishers, Netherlands, 2001.

50. ______, *Multihierarchical graph search*, IEEE Trans. on Pattern Analysis and Machine Intelligence **24** (2002), no. 1, 103–113.

51. R. Fikes and N. Nillson, *Strips: A new approach to the application of theorem proving to problem solving*, Artificial Intelligence **2** (1971), 189–208.
52. D. Floreano and F. Mondada, *Evolution of homing navigation in a real mobile robot*, IEEE Trans. on Systems, Man, and Cybernetics **26** (1996), 396–407.
53. T. W. Fong, C. Thorpe, and C. Baur, *Collaboration, dialogue, and human-robot interaction*, 10th International Symposium of Robotics Research, Lorne, Victoria, Australia, 2001, pp. 255 – 270.
54. T.W. Fong and C. Thorpe, *Robot as partner: Vehicle teleoperation with collaborative control*, In Proceedings of the NRL Workshop on Multi-Robot Systems, 2002, pp. 195–202.
55. J. Fritsch, M. Kleinehagenbrock, S. Lang, F. Loemker, G.A. Fink, and G. Sagerer, *Multi-model anchoring for human-robot-interaction*, Robotics and Autonomous System **43** (2003), no. 2-3, 133–147.
56. C. Galindo, J.A. Fernández-Madrigal, and J. González, *Hierarchical task planning through world abstraction*, IEEE Trans. on Robotics **20** (2004), no. 4, 667–690.
57. ______, *Self-adaptation of the symbolic world model of a mobile robot. and evolution-based approach*, 6th Int. FLINS Conf. on Applied Computational Intelligence, Blankenberge, Belgium, 2004, pp. 438–443.
58. C. Galindo, J. González, and J.A. Fernández-Madrigal, *Assistant robot, interactive task planning through multiple abstraction*, 16th European Conference on Artificial Intelligence, Valencia, Spain, 2004, pp. 1015–1016.
59. ______, *Interactive task planning in assistant robotics*, 5th Symposium on Intelligent Autonomous Vehicles, Lisbon, 2004.
60. C. Galindo, A. Saffiotti, S. Coradeschi, P. Buschka, J.A. Fernández-Madrigal, and J. González, *Multi-hierarchical semantic maps for mobile robotics*, IEEE/RSJ Int. Conf. on Intelligent Robots and Systems, Edmonton, Alberta (Canada), 2005, pp. 3492–3497.
61. M.R. Garey and D.S. Johnson, *Computers and intractability. a guide to the theory of np-completeness*, Freeman and Co. (eds.), 1979.
62. F. Giunchiglia and T. Walsh, *A theory of abstraction*, Artificial intelligence **57** (1992), no. 2-3, 323–389.
63. J. Goetz and S. Kiesler, *Cooperation with a robotic assistant*, CHI '02: Extended abstracts on Human factors in computing systems (New York, NY, USA), ACM Press, 2002, pp. 578–579.
64. J. González, C. Galindo, J.L. Blanco, A. Muoz, V. Arévalo, and J.A. Fernández-Madrigal, *The sena robotic wheelchair project*, ch. 20, pp. 235–250, DAAAM International Scientific Book 2006, Vienna, 2006.
65. J. González, A.J. Muoz, C. Galindo, J.A. Fernández-Madrigal, and J.L. Blanco, *A description of the sena robotic wheelchair*, Proc. of the 13th IEEE Mediterranean Electrotechnical Conference, May 16-19, Benalmdena, SPAIN, 2006.
66. J. González, A. Ollero, and A. Reina, *Map building for a mobile robot equipped with a laser range scanner*, IEEE Int. Conf. on Robotics and Automation, Leuven (Belgium), 1994.
67. J. Guivant, J. Nieto, F. Masson, and E. Nebot, *Navigation and mapping in large unstructured environments*, Int. Journal of Robotics Research **23** (2003), no. 4-5, 449–472.
68. S. Harnand, *Psychological and cognitive aspects of categorical perception: A critical overview*, Harnand S. (ed.), New York, Cambridge University Press, Chapter 1, 1987.

69. ______, *The symbol grounding problem*, Phys. D **42** (1990), no. 1-3, 335–346.
70. M. Henning and S. Vinoski, *Advanced corba programming with C++*, Addison-Wesley Professional, 1999.
71. S.C. Hirtle and J. Jonides, *Evidence of hierarchies in cognitive maps*, Memory and Cognition **13** (1985), no. 3, 208–217.
72. C.S. Holding, *Further evidence for the hierarchical representation of spatial information*, Journal of Environmental Psychology **14** (1994), 137–147.
73. I. Horswill, *Polly, a vision-based artificial agent*, The Proceedings of the Eleventh National Conference on Artificial Intelligence, Washington, DC, 1993, pp. 824–829.
74. F.F. Ingrand and E. Py, *An execution control system for autonomous robots*, IEEE Int. Conf. on Robotics and Automation, Washington, DC, 2002, pp. 1333–1338.
75. M. Isard and A. Blake, *Contour tracking by stochastic propagation of conditional density*, Proc. 4th European Conf. On Computer Vision, Cambridge,UK, 1996.
76. H. Jörg and N. Bernhard, *The ff planning system: Fast plan generation through heuristic search*, J. of Artificial Intelligence Research **14** (2001), 253–302.
77. L.P. Kaelbling, M.L. Littman, and A.R. Cassandra, *Planning and acting in partially observable stochastic domains*, Artificial Intelligence **101** (1998), 99–134.
78. K. Kawamura, R.A. Peters, C. Johnson, P. Nilas, and S. Thongchai, *Supervisory control of mobile robots using sensory egosphere*, IEEE Int. Symposium on Computational Intelligence in Robotics and Automation, Banff, Alberta, 2001, pp. 531–537.
79. R. Kennaway, *Graph Rewriting in Some Categories of Partial Morphisms*, Proc. 4th. Int. Workshop on Graph Grammars and their Application to Computer Science (Hartmut Ehrig, Hans-Jörg Kreowski, and Grzegorz Rozenberg, eds.), vol. 532, 1991, pp. 490–504.
80. O. Khatib, *Human-centered robotics and haptic interaction: From assistance to surgery, the emerging applications*, Third International Workshop on Robot Motion and Control, 2002, pp. 137–139.
81. M. Kleinehagenbrock, J. Fristch, and G. Sagerer, *Supporting advanced interaction capabilities on a mobile robot with a flexible control system*, IEEE/RSJ Int. Conf. on Intelligent Robots and Systems, Sendai, Japan, 2004, pp. 3649–3655.
82. C.A. Knoblock, *Generating abstraction hierarchies: An automated approach to reducing search in planning*, Kluwer Academic Publishers, 1993.
83. ______, *Automatically generating abstractions for planning*, Artificial Intelligence **68** (1994), no. 2, 243–302.
84. S. Koening, R. Goodwing, and R. Simmons, *Robot navigation with Markov models: A framework for path planning and learning with limited computational resources*, Lecture Notes in Artificial Intelligence **1903** (1996).
85. K. Konolige, *Large-scale map-making*, Proceedings of the Nineteenth National Conference on Artificial Intelligence, San Jose, CA, 2004, pp. 457–463.
86. K. Kouzoubov and D. Austin, *Hybrid Topological/Metric Approach to SLAM*, IEEE Int. Conf. on Robotics and Automation, New Orleans (LA), USA, 2004, pp. 872–877.
87. S. Kristensen, J. Klandt, F. Lohnert, and A. Stopp, *Human-friendly interaction for learning and cooperation*, IEEE Int. Conf. on Robotics and Automation, Seoul, Korea, 2001, pp. 2590–2595.

88. B.J. Kuipers, *Representing knowledge of large-scale space*, Ph.D. thesis, Massachusetts Institute of Technology, 1977.

89. ——, *The cognitive map: Could it have been any other way? Spatial orientation: Theory, research, and applications*, Picks H.L. and Acredolo L.P and New York, Plenum Press, 1983.

90. ——, *The spatial semantic hierarchy*, Artificial Intelligence **119** (2000), 191–233.

91. B.J. Kuipers and Y.T. Byun, *A qualitative approach to robot exploration and map-learning*, Workshop on Spatial Reasoning and Multi-Sensor Fusion, Charles, IL, 1987, pp. 390–404.

92. ——, *A robot exploration and mapping strategy based on a semantic hierarchy of spatial representations*, Robot Autonomous Syst. **8** (1991).

93. J.C. Latombe, *Robot motion planning*, Kluwer Academic Publishers, Boston, 1991.

94. S. Levine, D. Bell, L. Jaros, R. Simpson, Y. Koren, and J. Borenstein, *The navchair assistive wheelchair navigation system*, IEEE Transactions on Rehabilitation Engineering **7** (1999), no. 4, 443–451.

95. M. Lindstrom, A. Oreback, and H.I. Christensen, *Berra: A research architecture for service robots*, IEEE Int. Conf. on Robotics and Automation, San Francisco, CA, 2000, pp. 3278–3283.

96. B. Lisien, D. Morales, D. Silver, G. Kantor, I. Rekleitis, and H. Choset, *The hierarchical atlas*, IEEE Trans. on Robotics **21** (2005), no. 3, 473–481.

97. A. Loutfi, S. Coradeschi, and A. Saffiotti, *Maintaining coherent perceptual information using anchoring*, Proc. of the 19th IJCAI Conf. Edinburgh, UK, 2005.

98. F. Lu and E. Milios, *Robot pose estimation in unknown environments by matching 2d range scans*, Proc. IEEE Conf. on Computer Vision and Pattern Recognition, Seattle, WA, 1994, pp. 935–938.

99. ——, *Globally consistent range scan alignment for environments mapping*, Autonomous Robots **4** (1997), 333–349.

100. X. Lv and X. Huang, *Three-layered control architecture for microassembly with human-robot task plan interaction*, IEEE Int. Conf. on Robotics and Biomimetics, Shenyang, China, 2004, pp. 617–622.

101. H.F. Machiel, J.J. Wagner, N. Smaby, K. Chang, O. Madrigal, L.J: Leifer, and O. Khatib, *Provar assistive robot system architecture*, IEEE Int. Conf. on Robotics and Automation, Detroit, Michigan (USA), 1999, pp. 741–746.

102. D. Maio, D. Maltoni, and S. Rizzi, *Topological clustering of maps using a genetic algorithm*, Pattern Recognition Letters **16** (1995), 89–96.

103. J.L. Martínez, J. González, J. Morales, A. Mandow, S. Pedraza, and A. García-Cerezo, *Mobile robot motion estimation by 2d scan matching with genetic and iterative closest point algorithms*, Journal of Field Robotics **23** (2006), 21–34.

104. M. Mataric, *Synthesizing group behaviors*, Workshop on Dynamically Interacting Robots, IJCAI 1993, Chambery, France, 1993, pp. 1–10.

105. R.W. Matthews, *Insect behavior*, Matthews J.R., New York, Toronto, 1978.

106. T.P. McNamara, J.K Hardy, and S.C. Hirtle, *Subjective hierarchies in spatial memory*, Journal of Experimental Psychology: Learning, Memory, and Cognition **15** (1989), 211–227.

107. Sunrise Medical, *http://www.sunrisemedical.com*.

108. S. Minton, J.G. Carbonell, C.A. Knoblock, D. Kuokka, O. Etzioni, and Y. Gil, *Explanation-based learning: A problem solving perspective*, Artificial Intelligence **40** (1989), no. 1-3, 63–118.

109. Y. Mizoguchi, *A graph structure over the category of sets and partial functions*, Cahiers de topologie et geometrie differentielle categoriques **34** (1993), 2–12.

110. K. Morioka, J.H. Lee, and H. Hashimoto, *Human centered robotics in intelligent space*, In Proceedings of the IEEE International Conference on Robotics and Automation, Washington, DC, May 2002.

111. A.J. Muñoz and J. González, *2d landmark-based position estimation from a single image*, IEEE Int. Conf. on Robotics and Automation, Leuven, Belgium, 1998.

112. N. Nillson, *Shakey the robot*, Tech. report, No. 323, Artificial Intelligence Center, SRI International, Melon Park, CA, 1984.

113. P. Nordin, W. Bazhaf, and M. Brameier, *Evolution of a world model for a miniature robot using genetic programming*, Robotics and Autonomous Systems **25** (1998), 305–116.

114. S.P. Pari, R.S. Rao, S. Jungt, V. Kumar, J.P. Ostrowski, and C.J. Taylor, *Human robot interaction and usability studies for a smart wheelchair*, IEEE/RSJ Int. Conf. on Intelligent Robots and Systems, Las Vegas, Nevada, 2003, pp. 3206–3211.

115. I-P. Park and J.R. Kender, *Topological direction-giving and visual navigation in large environments*, Artificial Intelligence **78** (1995), 355–395.

116. A.L. Patalano, E.E. Smith, J. Jonides, and R.A. Koeppe, *Pet evidence for multiple strategies of categorization*, Cognitive, Affective, and Behavioral Neuroscience **1** (2001), no. 4, 360–370.

117. G.L. Peterson and D.J. Cook, *Incorporating decision-theoretic planning in a robot architecture*, Robotics and Autonomous Systems **42** (2003), no. 2, 89–106.

118. J. Pineau, N. Roy, and S. Thrun, *A hierarchical approach to pomdp planning and execution*, Workshop on Hierarchy and Memory in Reinforcement Learning (ICML), June 2001.

119. M. Pollack, S. Engberg, J.T. Matthews, S. Thrun, L. Brown, D. Colbry, C. Orosz, S. Ramakrishnan B. Peintner, J. Dunbar-Jacob, C. McCarthy, M. Montemerlo, J. Pineau, and N. Roy, *Pearl: A mobile robotic assistant for the elderly*, Workshop on Automation as Caregiver: the Role of Intelligent Technology in Elder Care, August 2002.

120. A. Poncela, E.J. Perez, A. Bandera, C. Urdiales, and F. Sandoval, *Efficient integration of metric and topological maps for directed exploration of unknown environments*, Robotics and Autonomous Systems **41** (2002), no. 1, 21–39.

121. A. Ram, R. Arkin, G. Boone, and M. Pearce, *Using genetic algorithms to learn reactive control parameters for autonomous robotic navigation*, Adaptive Behavior **2** (1994), no. 3, 277–304.

122. R.S. Rao, K. Conn, S.H. Jung, J. Katupitiya, T. Kientz, V. Kumar, J. Ostrowski, S. Patel, and C.J. Taylor, *Human robot interaction: Application to smart wheelchairs*, IEEE Int. Conf. on Robotics and Automation, Washington, DC, 2002, pp. 3583–3588.

123. C. R. Reeves and J. E. Rowe, *Genetic algorithms: Principles and perspectives. A guide to GA theory*, Kluwer Academic Publishers, 2003.

124. A. Reina and J. González, *A two-stage mobile robot localization method by overlapping segment-based maps*, Robotics and Autonomous Systems. Elsevier Science **31** (2000).

125. S. Rizzy, *A genetic approach to hierarchical clustering of euclidean graphs*, Int. Conf. on Pattern Recognition (ICPR'98) Brisbane, Australia, 1998, pp. 1543–1545.

126. K.A. Ross and C.R.B. Wright, *Discrete mathematics*, Prentice-Hall, New Jersey, 1992.

127. D.E. Rydeheard and R.M. Burstall, *Computational category theory*, Prentice Hall International (UK) Ltd, 1988.

128. E.D. Sacerdoti, *Planning in a hierarchy of abstraction spaces*, Artificial Intelligence **5** (1974), no. 2, 115–135.

129. A. Saffiotti, K. Konolige, and E.H. Ruspini, *A multivalued-logic approach to integrating planning and control*, Artificial Intelligence **76** (1995), 481–526.

130. P. Scerri, D. Pynadath, and M. Tambe, *Adjustable autonomy in real-world multi-agent environments*, AGENTS '01: Proceedings of the fifth international conference on autonomous agents, Montreal, Quebec (Canada), 2001, pp. 300–307.

131. J. Scholtz, *Theory and evaluation of human-robot interaction*, 36th International Conference on System Sciences, Hawai, 2003, pp. 1–10.

132. S. Shapiro, *Classical logic. The stanford encyclopedia of philosophy*, Edward N. Zalta eds., 2003.

133. T.B. Sheridan, *Telerobotics, automation and human supervisory control*, MIT Press,Cambridge, MA, 1992.

134. J. Shi and C. Tomasi, *Good features to track*, Proc. IEEE Conf. Comput. Vision and Pattern Recognition, 1994, pp. 593–600.

135. R. Simmons, R. Goodwin, K. Haigh, S. Koenig, and J. Sullivan, *A layered architecture for office delivery robots*, First International Conference on Autonomous Agents, February 1997, pp. 235 – 242.

136. J.M. Siskind, *Grounding language in perception*, Artificial Intelligence Review **8** (1995), 371–391.

137. M. Slack, *Navigation templates: mediating qualitative guidance and quantitative control in mobile robots*, IEEE Trans. on Systems, Man and Cybernetics **23** (1993), no. 2, 452–466.

138. D.E. Smith, J. Frank, and A.K. Jonsson, *Robot navigation with Markov models: A framework for path planning and learning with limited computational resources*, Lecture Notes in Artificial Intelligence **15** (2000), no. 1, 47–83.

139. E.M. Aogain S.O. Nuallain, P.Mc. Kevitt, *Two sciences of mind: Readings in cognitive science and consciousness. advances in consciousness research, v. 9*, John Benjamins Publishing Co, 1997.

140. L. Steels and R. Brooks, *The artificial life route to artificial intelligence*, Lawrence Erlbaum Associates, Inc., Publishers, 1995.

141. Z. Sun, D. Hsu, T. Jian, H. Kurniawati, and J.H. Reif, *Narrow passage sampling for probabilistic roadmap planning*, IEEE Trans. on Robotics **21** (2005), no. 6, 1105–1115.

142. G.J. Sussman, *A computer model of skill acquisition*, American Elsevier Publishing Company, 1975.

143. K.A. Tahboub, *A semi-autonomous reactive control architecture*, Intelligent Robotics Systems **32** (2001), no. 4, 445–459.

144. I. Takeuchi and T. Furuhashi, *Self-organization of grounded symbols for fusions of symbolic processing and parallel distribted processing*, Proc. of 1998 IEEE Int. Conf. on Fuzzy Systems, 1998, pp. 715–720.

145. J. Tani, *Model based learning for mobile robot navigation from the dynamical systems perspective*, IEEE Trans. System Man and Cybernetics Part B **26** (1996), no. 3, 421–436.

146. T. Taniguchi and T. Sawaragi, *Self-organization of inner symbols for chase: symbol organization and embodiment*, Conf on Systems, Man and Cybernetics, 2004, 2004, pp. 2073–2078.

147. S. Thrun, *Robotic mapping: A survey*, Exploring Artificial Intelligence in the New Millennium, Morgan Kaufmann, 2002.

148. S. Thrun and Bücken, *Integrating grid-based and topological maps for mobile robot navigation*, 13th National Conf. on Artificial Intelligence, Portland, Oregon, 1996, pp. 944–951.

149. S. Thrun, D. Fox, and W. Burgard, *Probabilistic mapping of an environment by a mobile robot*, Proc. IEEE Int. Conf. on Robotics and Automation, 1998.

150. Text to Speech Software 2nd Speech Center, *http://www.zero2000.com*.

151. M.R. Trudeau, *Introduction to graph theory*, Dover Publications, 1993.

152. C. Urdiales, E.J. Perez, F. Sandoval, and J. Vazquez-Salceda, *A hybrid architecture for autonomous navigation in dynamic environments*, IEEE/WIC International Conference on Intelligent Agent Technology (IAT 2003), Halifax, Canada, 2003, pp. 225–232.

153. M. Veloso, J. Carbonell, A. Perez, D. Borrajo, E. Fink, and J. Blythe, *Integrating planning and learning: The PRODIGY architecture*, Journal of Experimental and Theoretical Artificial Intelligence **7** (1995), no. 1, 81–120.

154. IBM ViaVoice, *http://www-3.ibm.com/software/voice/viavoice/*.

155. B. Wah, M. Lowrie, and G. Li, *Computers for symbolic processing*, Proc. of the IEEE (77),4, 1989, pp. 509–540.

156. J.O. Wallgrün, *Hierarchical Voronoi-based route graph representations for planning, spatial reasoning, and communication*, 4th International Cognitive Robotics Workshop, Valencia, España, 2004.

157. G. Wasson, J. Gunderson, S. Graves, and R. Felder, *An assistive robotic agent for pedestrian mobility*, AGENTS '01: Proceedings of the fifth international conference on Autonomous agents, 2001, pp. 169–173.

158. D. Whitley, *A genetic algorithm tutorial*, Statistics and Computing **4** (1994), 65–85.

159. Q. Yang, *Intelligent planning: A decomposition and abstraction based approach*, Springer Verlag, 1997.

160. U.R. Zimmer, *Robust world-modeling and navigation in a real world*, In Neurocomputing, Special Issue, 1996.

161. M. Zweben and M. Fox, *Intelligent scheduling*, M. Kaufmann, 1994.

# Index

GPSR Compliance
The European Union's (EU) General Product Safety Regulation (GPSR) is a set
of rules that requires consumer products to be safe and our obligations to
ensure this.

If you have any concerns about our products, you can contact us on

ProductSafety@springernature.com

In case Publisher is established outside the EU, the EU authorized
representative is:

Springer Nature Customer Service Center GmbH
Europaplatz 3
69115 Heidelberg, Germany